BANKING AND BANKING DEVELOPMENTS

BANKING

PERFORMANCE, CHALLENGES AND PROSPECTS FOR DEVELOPMENT

BANKING AND BANKING DEVELOPMENTS

Additional books in this series can be found on Nova's website under the Series tab.

Additional E-books in this series can be found on Nova's website under the E-book tab.

BANKING AND BANKING DEVELOPMENTS

BANKING

PERFORMANCE, CHALLENGES AND PROSPECTS FOR DEVELOPMENT

JANICE P. SIMMONS
EDITOR

NOVINKA

New York

Library of Congress Cataloging-in-Publication Data

Banking : performance, challenges and prospects for development / editor, Janice P. Simmons.
pages cm
Includes index.
ISBN 978-1-62948-897-4 (soft cover)
1. Banks and banking--Europe. 2. Multilevel models (Statistics) I. Simmons, Janice P., editor of compilation.
HG1615.B356 2014
332.1094--dc23

2013047182

Published by Nova Science Publishers, Inc. † New York

CONTENTS

PREFACE

There is ample literature that argues commercial banks within a country and across countries are in economic harmony when it comes to their capital/asset ratio response to changes in such economic determinants as cost, loans, and profitability. This book focuses on answering the question if European banks are in economic harmony. It continues with discussing banking employment in the early twenty-first century and ends with a national cohort study on posttraumatic stress symptoms in individuals following a bank robbery.

Chapter 1 – There is ample literature that argues commercial banks within a country and across countries are in economic harmony when it comes to their capital/asset ratio response to changes in such economic determinants as cost, loans, and profitability. The main reason to expect such harmony is the fact that banks are heavily supervised and regulated. This chapter does not dispute the literature. What are missing from the literature are quantitative or statistical studies presenting evidence of economic harmonization. A reduced-form equation relating the capital account ratio to several micro and macro variables, particularly the profitability variable, for a limited sample of commercial banks in nine European countries over eleven years, 1991-2001, was constructed. What is unique about the statistical method is the fact that the Hierarchical Linear Model (HLM) approach was used to test the harmonization hypothesis. The statistical results indicated that while some random differences in bank behavior as indicated by the random intercept and slope deviations within and across countries and over the years did exist, by and large, most of the differences or deviations from the fixed-effects means were not significantly different from zero. The harmonization hypothesis was accepted and the results were consistent with the literature. What is novel about this chapter is that now there is hard statistical evidence of harmonization. European banks' economic behavior gave evidence of being in harmony within and over countries and years. Some policy implications are discussed briefly.

Chapter 2 – The effects of consolidation in the commercial banking sector on employment are examined using data from the SNL's banking database, as well as Bureau of Labor Statistics data on employment. This paper looks at the effects of the number of banks on employment. It is found that there is a negative relationship between consolidation in the banking industry and banking employment.

Chapter 3 – Each year, numerous bank robberies take place worldwide. Even so, only few studies have investigated the psychological sequelae of bank robbery and little is known about the risk factors associated with the development of posttraumatic stress disorder (PTSD) following this potentially traumatic event. Knowledge about risk factors related to PTSD may allow for preventive measures to be taken against the development of PTSD and reduce the large cost associated with the disorder. Furthermore, the few existing studies are characterized by several limitations such as the use of small convenience or self-selecting samples. To overcome these limitations, the authors investigated the estimated prevalence rate of acute stress disorder (ASD) and PTSD and predictors of PTSD severity in a Danish cohort study of all bank employees exposed to robbery during one year (N = 614). A total of 450 employees (73 %) filled out the first questionnaire a week after the robbery (T1). Of these, 371 employees (82 %) filled out the second questionnaire six months after the robbery (T2). Results showed that 11.1% of the participants suffered from ASD (T1) and 6.2 % suffered from PTSD (T2). The results of a hierarchical regression analysis showed that 51 % of the variance in PTSD severity could be explained with only peritraumatic dissociation, ASD severity, and negative cognitions about self being significant predictors. However, interpretation of the role of peritraumatic dissociation in the regression model was hampered by a statistical artifact (negative suppression). The strongest predictors of PTSD severity were ASD severity followed by negative cognitions about self.

This indicates that victims with high levels of ASD are at increased risk of developing PTSD following bank robbery. Thus, screening for PTSD following bank robbery should focus on ASD severity. In accordance with a cognitive model of PTSD and the DSM-5, the results show that posttraumatic negative cognitions about self also seem to play a pivotal role in the development of PTSD following bank robbery.

This indicates that it may be possible to develop preventive cognitive interventions targeted at bank employees that focus on changing negative cognitions about self.

In: Banking
Editor: Janice P. Simmons
ISBN: 978-1-62948-897-4

Chapter 1

ARE EUROPEAN BANKS IN ECONOMIC HARMONY? A HLM APPROACH

James P. Gander[1*]
[1]Economics Department, University of Utah, Salt Lake City, UT, US

ABSTRACT

There is ample literature that argues commercial banks within a country and across countries are in economic harmony when it comes to their capital/asset ratio response to changes in such economic determinants as cost, loans, and profitability. The main reason to expect such harmony is the fact that banks are heavily supervised and regulated. This chapter does not dispute the literature. What are missing from the literature are quantitative or statistical studies presenting evidence of economic harmonization. A reduced-form equation relating the capital account ratio to several micro and macro variables, particularly the profitability variable, for a limited sample of commercial banks in nine European countries over eleven years, 1991-2001, was constructed. What is unique about the statistical method is the fact that the Hierarchical Linear Model (HLM) approach was used to test the harmonization hypothesis. The statistical results indicated that while some random differences in bank behavior as indicated by the random intercept and slope deviations within and across countries and over the years did exist, by and large, most of the differences or deviations from the fixed-effects

* 801-581-7481, Fax, 801-585-5649, gander@economics.utah.edu.

means were not significantly different from zero. The harmonization hypothesis was accepted and the results were consistent with the literature. What is novel about this chapter is that now there is hard statistical evidence of harmonization. European banks' economic behavior gave evidence of being in harmony within and over countries and years. Some policy implications are discussed briefly.

Keywords: Bank behavior; Profit; Capital account ratios, Harmonization

JEL classification: C23; C40; C51; G21; G28

INTRODUCTION

The purpose of this chapter is to test the proposition or hypothesis, Are European banks in economic harmony with respect to the behavior of their capital accounts? In other words, do they all responds the same way? To be more specific, by economic harmony is meant that banks respond similarly to given economic conditions, internal to the bank and external to the bank. Perhaps such harmony is to be expected, since as Fama (1985) argues that banks are special in that they are subject to closer regulation and supervision (see, for example, Levine, 2004; Barth et al., 2004; and Barth et al., 2006) than other types of firms. Although this does not mean that other types of firms are not subject to close supervision (for example, meat processing firms and steel mills are subject to government inspection). It is just that banks have a special role in a country's monetary system (Corrigan, 1982).

So, why should economic harmony be an issue? Conceptually, it is not. But, one answer is that statistically there is enough variability in banks' responses, in spite of regulations, to suggest that the responses of banks within and between countries may be significantly different. It is the possibility of this difference that is being investigated and tested. Regardless of the specifics about regulation, the issue is what do the facts suggest? There are no quantitative or statistical studies that examine that question. As the literature shows, there is a great amount of discussion and policy rhetoric on bank supervision and regulation and these imply a certain degree of harmonization, but the purpose of this paper is not to argue what is best in terms of regulation, but to present a concrete statistical test of harmonization.

If the harmonization statistical test is accepted (as the literature suggests it will be), one interesting implication for the organization or structure of the banking industry relates to the number of banks within a country and across countries. If all individual banks are in harmony in terms of their capital accounts within and between countries, this suggests that one large integrated bank can perform as efficiently as a market collection of individual banks. This suggestion is based on the proposition that the internalization of many individual decisions under one entity will, under certain conditions, give results equivalent to the aggregate of the individual decisions. From a biological perspective, the body and the cells therein are in harmony. So are the bee hive and the bees. All the solders march to the same step as the army. In other words, harmonization among banks suggests that one large bank is equivalent. This proposition follows closely Coase's (1937) theorem on internal versus external resource allocation and firm size.

The harmonization statistical test involves using the Hierarchical Linear Model (HLM) with nine European countries' commercial banks over the years 1991-2001. Most of the countries are members of the European Union (EU). They represent a mix of large and small economies. The economies are quite diverse in terms of economic structure and policy, which if anything would suggest a lack of harmony. The time period chosen for the test was one of convenience in that data were available. It also can be argued that it was one of relatively stable economic conditions, compared to the unstable financial conditions presently in 2011-2012.

There is a decided effort today for banks to increase their capital accounts (see, WSJ, November 11, 2011), which if successful will result in a certain degree of harmony. But, the purpose of this chapter, as indicated before, is to bring a quantitative perspective to the notion of harmony. This is not to say that the presence of the EU and the Basel capital ratio guide lines do not have a role in creating such harmony. At the very least, the Basel accord sets a minimum capital ratio requirement for country regulators to follow, if they so choose. The actual capital ratios are quite often more than the minimum guidelines, resulting in what is referred to as a buffer (see, for example, Jokipii and Milne, 2007). Further, this chapter takes a quantitative approach to harmony (for a non-quantitative approach, see, Singer, 2004).

Finally, while there is literature on the use of various quantitative methods to analyze European banks and their capital accounts (for example, Jokipii and Milne, 2007 and Matejasak, Teply, and Cernohorsky, 2009), as stated earlier, there appears to be no literature on using the HLM quantitative method for bank capital account analysis. Dong and Stettler (2011) have used the HLM approach to analyze corporate disclosure scores, but not to analyze banks' capital account ratios.

In general, the HLM approach is ideal for testing the harmonization hypothesis. The key question is, Do banks in a given country behave more similar in terms of their capital accounts than do banks in another country? Rather than run a standard cross-sectional time series regression with fixed-effects dummy variables for the nine countries and the eleven years, the HLM approach recognizes that the banks are nested within a country and that there may be within-country random effects and across-countries random effects and over-the-years random effects, in addition to fixed effects. A reduced-form equation is developed (the details will be given momentarily) with the capital account ratio as the dependent variable and micro and micro*macro interaction variables as independent variables.

The focus is on the profitability variable for the banks and whether and to what extent the intercepts and the slopes (of profitability) vary randomly within and across countries and over the years. The harmonization test involves determining whether or not the random effect differences in the intercepts and slopes are significantly different from zero or not. The harmonization hypothesis implies they are not. To use an analogy, smart students are good academic performers regardless of the course, the school, or the district. Banks behave harmoniously regardless of the country or time period.

In what follows, the reduced-form equation is developed in the next section. Then, the HLM is applied to the reduced-form equation in the third section. In the fourth section, the sample and data are discussed and the statistical results are analyzed. The last section contains a summary and conclusions.

REDUCED-FORM EQUATION

The reduced-form equation is essentially a relationship between the bank's capital adequacy or account ratio and its micro and macro determinants, especially the micro profitability variable (or ratio) and a proxy for risk. The equation is in the spirit of those used in the literature relating bank efficiency, capital, and risk (see, for example, equation 4 in Fiordelisi, Marques-Ibanez, and Molyneux, 2011 and the literature cited there in). In symbolic terms the equation here is given by

$$LOGCAPASTt_{ijt} = a + b1{*}RGDPC + b2{*}NPROFEQ + b3{*}(NPROFEQ{*}RGDPC) + b4{*}ADMREV + b5{*}INVCUSTAST + e, \quad (1)$$

where the LOGCAPAST is the log of the capital account to assets ratio for bank "i" in country "j" and year "t" (subscripts are left off the regressors for simplicity). The capital account can have several different definitions (for example, the tiers in the Basel definition), but the one used here is based on Standard & Poor's definition. It consists of the sum of long-term debt plus minority stock interests plus shareholder equity which includes reserves and retained earnings. This wider definition has the least number of missing values and is, therefore, more suitable for the statistical analysis in this chapter. All of the capital components provide the bank with financial funds to hedge against the risk of default of primarily short-term customer loans (to households and businesses) and enable the bank to increase its capital ratio as needed.

The RGDPC is real gross domestic product per capita in U.S. dollars (base year is 1995) for each country and year. The RGDPC macro variable is used as an economic condition indicator. It is expected to be negatively related to the capital account ratio, although the literature is mixed on the relationship (see, for example, Jokipii and Milne, 2007). The NPROFEQ is the bank's profitability variable (net income after taxes and appropriations to untaxed reserves in the balance sheet relative to shareholder equity). All other things being equal, it is expected to be positively related to the capital account by virtue of its effect on retained earnings. The ADMREV is the ratio of total staff and administrative personnel and equipment and lease expenses to total bank revenue (interest received, fees, and returns on market investments). This variable is used, in effect, as an efficiency index to capture the effect of bank loan-making costs on the capital account. It is expected to be inversely related to the capital account, since as a cost, it is inversely related to net income. The (NEPROFEQ*RGDPC) is a micro*macro interaction variable and is expected to be positively related to the capital account.

The INVCUSTAST variable is the inverse of the ratio of customer loans (recorded as net of the reserves for credit losses) to total assets. It is used as a risk indicator. The capital, efficiency (profit), and risk variables used here while not identical are similar to those used by Fiordelisi, Marques-Ibanez, and Molyneux (2011), as indicated earlier. The inverse ratio is used rather than the direct ratio of customer loans to assets, for its coefficient was significant. The use is based on the assumption that the lower the inverse ratio, the higher the direct ratio (more loans are leveraged per assets), but since the loans are net of the loss reserves, risk is relatively less, so the capital ratio can be lower. Thus, the inverse ratio is expected to be positively related to the bank's capital account ratio. The e is the i, j, and t residual. The variables are all ratios (with the exception of RGDPC which is in real U.S. dollars) to avoid having to convert local currency units to a common currency unit. Equation (1) is essentially the fixed-effect component of the HLM equation given next.

THE HLM EQUATION

A brief description of the HLM approach follows. In general terms, the HLM equation can be given by (following SAS's notation):

$$y = X\beta + Zu + e, \quad (2)$$

where u is the vector of random coefficients, Z is the design matrix (here, the NPROFEQ variable by country and by year plus the intercept by country for all years pooled), and the $X\beta + e$ is the fixed-effect part, equation (1). The countries are referred to as the subjects and the years as the groups. The banks are the units of observation. Specifically, for the present analysis, the expanded equation (1) is given by

$$\text{Equation (1) plus } u1_{jt} * NPROFEQ_{ijt} + u_{oj}, \quad (3)$$

where u_{oj} is the random intercept for the country-specific equation for all years pooled and $u1_{jt}$ is the random slope effects by country and by year. The two class levels are country (values 1 to 9) and year (values 1 to 11). Both classes are used for the random slopes, but only the country level is used for the random intercepts. Econometrically, this design worked best.

Since the focus is on the bank's profitability variable, NPROFEQ, it can vary across banks within a given country for a given year and across years for a given bank and country. These variances are at the core of the harmonization hypothesis. The banks are unique to a given country and identified by their gvkey (global vantage key). The fixed-effects equation estimates are the mean of the country-level intercepts and the means of the slopes for the X-variables for the country level equations. The intercepts and the slopes for each country-level equation can be compared for their random differences, to test whether they are significantly different from zero. It should be noted that the individual banks in a given country cannot be compared to the individual banks in another country. Only the means or averages can be compared. It should also be noted that the u's and e's are assumed to be normally distributed. A test on the LOGCAPAST variable showed it to be normally distributed.

The Sample Data and Statistical Results

The bank micro data are from the Standard & Poor's Compustat, Global Vantage (GV) file (2002) covering the period 1991-2001. The data are unbalanced panel data across time and across countries. The initial sample size was 1780 observations, which after accounting for missing values was reduced to 1641 observations over nine countries, eleven years, and some 252 banks. The nine countries are Portugal, Turkey, Germany, Spain, France, Italy, Austria, Greece, and Switzerland. The data are taken from the balance sheets and income statements of publicly trade depository commercial banks in Europe. For some of the predictor variables suggested by bank theory, see, for example, Allen and Santomero (1997); Freixas and Rochet (1997); Barnes (1987); Damodaran (1997); and Kallunki, Martikainen, and Perttunen (1996).

The macro data are from the World Bank's World Development Indicators file. The macro variables considered in the initial analysis were real GDP (base year 1995) per capita in U.S. dollars (RGDPC), its rate of growth, the ratio of M2 money supply to GDP, and the rate of growth of the M2 money supply. Real GDP per capita was in the final analysis the best macroeconomic indicator. As indicated at the outset, the macro variable is used as a stand-alone variable and also as a micro*macro interactive variable. If there is a perception of "good economic times," as indicated by the macro variable, then the bank may be more comfortable with a smaller capital adequacy ratio (given any legal requirement). If the opposite is the case, then a larger ratio may be preferred. Since bank assets are largely customers' loans, then variation in these will essentially affect the capital adequacy ratio.

The statistical results are recorded in Table 1. The fixed-effects coefficients are all significant at p = .004 or less. The positive sign for the RGDPC coefficient was unexpected, since a larger RGDPC implies an economically healthier economy and, therefore, less risk of loan defaults. As such, a lower capital account ratio should be expected. As indicated earlier, the literature is mixed on the direction of this relationship. The negative sign for the interaction effect is expected for the reason given before for the RGDPC effect. The coefficient for NPROFEQ is positive as expected, since profit adds to the capital account as defined earlier. The negative sign for the ADMREV cost coefficient implies that as the cost of loan making increases, the capital account ratio will fall due to the effect of cost on the bank's net income (as explained earlier). The positive sign for the INVCUSTAST coefficient is expected (as explained earlier).

Table 1. HLM Statistical Results for European Banks

Dep. Var.:	LOGCAPAST		
Solution For Fixed Effects			
Coefficient	**Mean Estimate**	**t-Value**	**p-Value**
CONSTANT	-2.719	-10.05	<.0001
RGDPC	.000056	5.71	<.0001
NPROFEQ	1.275	3.13	.0018
NPRO*RGDPC	-.00004	-2.75	.0001
ADMREV	-1.152	-10.66	<.0001

Solution for Random Effects

Coefficient	**Ctryinc**	**Year**	**Mean Deviation**	**t-Value**	**p-Value**
CONSTANT	AUT	All	0.0983	0.52	0.604
NPROFEQ	AUT	1991	0.1924	0.07	0.946
CONSTANT	CHE	All	-0.8158	-2.97	0.003
NPROFEQ	CHE	1991	0.2917	0.22	0.826
CONSTANT	DEU	All	0.0715	0.38	0.705
NPROFEQ	DEU	1991	3.295	2.04	0.042
CONSTANT	ESP	All	-0.174	-0.96	0.335
NPROFEQ	ESP	1991	-1.506	-2.09	0.037
CONSTANT	FRA	All	-0.319	-1.81	0.070
NPROFEQ	FRA	1991	1.382	5.60	0.0001
CONSTANT	GRC	All	-0.126	-0.62	0.537
NPROFEQ	GRC	1991	0.000	0.000	1.000
CONSTANT	ITA	All	0.297	1.76	0.079
NPROFEQ	ITA	1991	-6.368	-5.50	0.0001
CONSTANT	PRT	All	0.060	0.30	0.766
NPROFEQ	PRT	1991	0.000	0.000	1.000
CONSTANT	TUR	All	0.907	3.41	0.001
NPROFEQ	TUR	1991	0.000	0.000	1.000

Notes: The mean coefficient estimates in the FE part of the table are averages across all countries and years and are all statistically significant. In the RE part of the table, the estimates are deviations from the FE means. The nine deviations for the country-specific constants (intercepts) sum to zero as expected. The NPROFEQ slope coefficients by country are only recorded for the year 1991 as an example, to conserve space. Of the nine intercept coefficients, each covering all eleven years, four are significant at p <=10 percent, only two at p <=5 percent. Of the 99 slope deviations (nine countries and eleven years), only 17 are significant at p

<=10 percent. Of the 13 covariance parameter estimates (not recorded here), one for the overall intercept, eleven for the NPROFEQ slopes that vary across the nine countries for each year, and one for the residual, six are significant at the p <= 10 percent. Of these six, five are not significant when p <= 5 percent is used. For each year, the covariance of the slopes across the nine countries in each year is an indication of the harmonization hypothesis. The sample size was 1641 observations. The method of estimation was the restricted maximum likelihood method (REML) using SAS.

The main interest of the analysis of bank behavior relates to the harmony hypothesis as explained at the outset. The conclusion relevant to this hypothesis is suggested by the random effects (RE) results in the second part of Table 1. The second part records the deviations of the random coefficients from the fixed-effects coefficients which are the overall means coefficients. As indicated in the notes to Table 1, a small proportion (about 19.4 percent) of these deviations was statistically significantly different from zero at p =< 10 percent. Out of 108 random deviations (nine country intercept deviations and 99 NPROFEQ random-coefficient deviations over time and country), only 21 were significantly different from zero at p =< 10 percent. The table shows only 18 of these deviations (for nine constants over all the years and nine slopes for 1991 only) to save space (the author can e-mail the full set of deviations upon request).

In general, the conclusion is that for all banks in all nine countries over eleven years, the random deviations for all the intercepts and profit rate slopes are not significantly different from zero. Hence, banks very much all behave the same way in terms of their capital account-profit rate relationship, regardless of the country or year. The harmony hypothesis is, therefore, accepted. Theoretically, one could argue that this is the way a European-wide competitive market should be expected to behave, but it can also be argued that an integrated monopoly structure would also behave this way. Given the large number of banks in this analysis (252), the former argument seems more appropriate.

CONCLUSION

A reduced-form equation relating the log of the capital account ratio to several micro and macro variables, particularly the profitability variable, for the commercial banks in nine European countries over eleven years, 1991-

2001, was constructed. The assumption being that banks realized profits add to their capital account ratios. The equation consisted of a fixed-effects part and a random-effects part. The HLM approach was used to test the harmonization hypothesis relating the capital account ratio to the profit rate across the countries and over the years. The harmonization implies that all banks behave consistently and collectively as if one bank.

The statistical results indicated that while some differences in bank behavior as indicated by the random intercept and slope deviations across countries and over years did exist, by and large, most of the random differences or deviations from the fixed-effects means were not significantly different from zero. The harmonization hypothesis was accepted. European bank behavior gave evidence of being in harmony and uniform over countries and years. They collectively behave as if there were one large bank.

In terms of present-day economic conditions in Europe and particularly in terms of the commercial banks behavior therein, harmonization is not only a desirable policy goal, but it is also one that is achievable. More research on the policy implications of the harmonization thesis is needed in the future. Suffice it to say at this time that if harmonization is a natural market determined phenomenon, as the evidence in this paper seems to suggest, then regulatory policy leaders should act as a catalyst to ensure that the harmonization process is realized (see, for a test of banks' positive responses to regulatory requirements (pressure), Matejasak, Teply, and Cernohorsky, 2009). While one would expect harmonization to be equivalent to a monopoly-type banking industry structure, the fact that it occurs under market conditions (given the regulatory regime), suggest that monopoly is not needed and for other reason should be avoided.

The evidence in the present chapter does not imply that a-specific, country-level regulatory policy will not have a pro-active role in influencing the country's bank behavior, particularly in terms of banks' capital ratios. But, as is generally apparent from the news media, one-by-one each European country (at least some of those in this chapter) is responding the same way (by raising capital requirements, among other policy moves) to the European bank (and government) financial crisis, thus creating, in effect, harmonization (as Matejasak, et al., 2009 found).

REFERENCES

Allen, F. & Santomero, A. (1997). The Theory of Financial Intermediation. *Journal of Banking & Finance*, 21 (11-12), 1461-1485.

Barnes, P. (1987). The Analysis and Use of Financial Ratios: A Review Article. *Journal of Business, Finance & Accounting*, 14 (4), 449-461.

Barth, J. R., Caprio, G, Jr. & Levine, R. (2004). Bank regulation and supervision: what works best? *Journal of Financial Intermediation*, 13 (April), 205-248.

———. (2006). Rethinking Bank Supervision and Regulation: Until Angels Govern. Cambridge: Cambridge University Press.

Coase, R. (1937). The Nature of the Firm. *Economica*, 4 (November), 386-405.

Corrigan, E. (1982). Are banks special? Annual Report. Federal Reserve Bank of Minneapolis.

Damodaran, A. (1997). Corporate Finance: Theory and Practice. New York: John Wiley & Sons.

Dong, M. & Stettler, A. (2011). Estimating firm-level and country-level effects in Cross-sectional analyses: An application of hierarchical modeling in corporate Disclosure studies. *The International Journal of Accounting*, 46 (3), 271-303.

Fama, z. (1985). What's different about banks? *Journal of Monetary Economics*, 15(1), 29-39.

Fiordelisi, F. & Molyneux, P. (2010). The Determinants of Shareholder Value in European Banking, *Journal of Banking and Finance*, 34, 1189-1200.

Fiordelisi, F., Marques-Ibanez, D., & Molyneux, P. (2011). Efficiency and Risk in European Banking. *Journal of Banking and Finance,* 35, 1315-1326.

Freixas, X. & Rochet, J.-C. (1997). Microeconomics of Banking. Cambridge, Massachusetts: The MIT Press.

Jokipii, T. & Milne, A. (2007). The Cyclical Behaviour of European Bank Capital Buffers. *The Swedish Institute for Financial Research,* No. 56 (July), 1-22.

Kallunki, J.-P., Martikainen, T. & Perttunen, J. (1996). The Proportionality of Financial Ratios: Implications for Ratio Classifications. *Applied Financial Economics,* 6, 535-41.

Levine, R. (2004). The Corporate Governance of Banks—a Concise Discussion of Concepts and Evidence. *Policy Research Working Paper Series,* 3404, The World Bank.

Matejasak, M., Teply, P. & Cernohorsky, J. (2009). The Impact of Regulation of Banks in the US and the EU-15. *Ekonomie a Management*, 12 (3), 58-68.

Singer, D. A. (2004). Capital Rules: The Domestic Politics of International Regulatory Harmonization. *International Organization*, 58 (3, Summer), 531-565.

Standard & Poor's (2002). Compustat, Global Vantage File. McGraw-Hill, Inc.

Wall Street Journal (2011). "Heard on the Street: Europe's Drama is Far From Over," Comment #1, November 11.

World Bank Group. World Development Indicators database, CD-ROM, August 2003.

In: Banking
Editor: Janice P. Simmons
ISBN: 978-1-62948-897-4

Chapter 2

BANKING EMPLOYMENT IN THE EARLY TWENTY-FIRST CENTURY: REGIONAL EFFECTS OF CONSOLIDATION

Russell Kashian and Derrek Grunfelder-McCrank
University of Wisconsin, Whitewater, WI, US

ABSTRACT

The effects of consolidation in the commercial banking sector on employment are examined using data from the SNL's banking database, as well as Bureau of Labor Statistics data on employment. This paper looks at the effects of the number of banks on employment. It is found that there is a negative relationship between consolidation in the banking industry and banking employment.

INTRODUCTION

One of the outcomes of the recent financial turmoil has been continued consolidation in the banking sector. While "too big to fail" has become a negative rallying call against the large banks and government intervention, banks continue to consolidate and get larger. One question open to research is whether this consolidation leads to efficiencies in regard to employment. If so, then there will be a decrease in employment along with an increase in average assets and decrease in the number of banks. One tool to begin answering this question is the shift share method, which can parcel out regional impacts. In recent years, economists have used the Shift-share analysis as a base from which to build further analyses. This study builds upon that approach by combining shift-share and regression techniques as found in in Blien and Wolf (2002) as well as Berzeg (1998).

The question at hand is how has commercial banking employment been rebalanced in the United States over the period from 2000-2012. Given that this period includes the impact of the bursting of the "dot-com" bubble, the impact of Y2K and the Financial Crisis of 2008, the changing structure of regional employment is ripe for review. Regional disparities in banking employment have led to the question of whether certain areas in the U.S. were oversaturated with banking employment at the beginning of this period. Examining commercial banking employment by regions of the United States as defined by the Bureau of Economic Analysis allows any redistribution to be identified.

Given there is an observed redistribution of jobs in the commercial banking industry, this paper uses ordinary least squares to expand the analysis. Several of the factors considered in regards to commercial banking employment are assets, institutions, branches, population, and the region. These are also variables used in the earlier literature on shift share analysis.

Criticisms of Shift-Share

The shift-share method can examine the changing competitiveness of the banking sector to calculate the forces affecting their growth or decline. There are three sources that create these impacts. The first determines the local change (growth or decline) associated with any national change in employment. The second reviews the local changes brought about by the local concentration of businesses (specifically banking). The third is the local change that evolves out of any competitive advantage (or disadvantage) local firms may have relative to the nation. While shift-share analysis offers insight into the magnitude of these factors, it is simply a descriptive tool. An additional econometric analysis is included that examines some of the determinants of employment changes.

One of the criticisms of the shift-share techniques is that they are a static, partial equilibrium data description tool that employs non-parametric testing. These models do not allow for any intermediate point, so the tool ignores changes in both the structure of the sector under review and in the levels of regional employment. Much of this critique is included in Mayor, Lopez, Perez (2007) and

Chilian (2012)[1]. Since considering the initial year when weighting, Stilwell (1969) offers the concept of using weights based on the final year to address this issue. Additionally, there are those who simply believe that some combination of the initial year and the final year is a better gauge to use.

A second criticism is that the static shift tool does not take into account continuous changes in the size of regional employment. This shift can affect the computed effects. This "compounding effect" (Barff and Knight, 1998) refers to the phenomenon that results when the analysis produces incorrectly weighted national growth numbers. This result feeds through the process resulting in an incorrect result when regional employment growth outperforms (or underperforms) the national average. A dynamic shift share analysis corrects for this issue. The shift-share is made dynamic by splitting the data into two or more databases. This essentially breaks the results down into various sub-time periods within the overarching time horizon.

By breaking the data into pre- and post-financial crisis epochs, we address both of these criticisms. It confronts the first criticism by eliminating the idea of an arbitrary start and end period. We begin at the apex of "dot-com" bubble, break the data up with the advent of the financial crisis and finish with the current period. As will be noted in Table 3, both epochs found similar results regarding the "competitive" growth effect. So we effectively counter both of the earlier criticisms by employing a dynamic shift-share model.

EMPLOYMENT TRENDS

The primary sources of data used are the *Statistics on Depository Institutions* data set from the FDIC as well as the *Quarterly Census of Employment and Wages* published by the Bureau of Labor Statistics. These data covers the time period from 2000-2012. This approach is designed to capture the recession of 2001, the expansion that followed, and the "Great Recession" (as the most recent recession is being called). The Regions used come from the Bureau of Economic Analysis, and are used to examine trends in banking employment at the regional level.

Employment in the commercial banking sector (hence referred to simply as banking) rose nationally by 5.14% from 2001-2012. This is compared with the contraction of 8.6% in the Mideast. This is important to note, as the Mideast is

[1] A good summation of criticisms of shift share analysis can be found in Mayor, Lopez, Perez (2007), and in Chilian (2012)

home to the city of New York, which is a major financial hub. Compare this to the 27.18% increase in the Southwest or the 11.86% expansion in banking employment in the Far West.

Table 1. Changes in Employment 2001-2012

Region	Change in Banking Employment (Percent)	Change in Total Employment (Percent)
New England	-12.42	-1.73
Mideast	-8.60	0.77
Great Lakes	-5.99	-6.17
Plains	16.95	2.08
Southeast	7.04	0.67
Southwest	27.18	12.43
Rocky Mountain	13.61	7.13
Far West	11.86	1.99

Table 2 shows a surprising trend. A reduction in the number of banks coincides with an increasing number of branches in a region. This is shown for every region from 2001-2012 except for the Great Lakes region, which bucks the trend and has the number of branches decrease along with the number of banks. This finding suggests that consolidation is positively correlated with an increase in the number of branches.

Table 2. Changes in Banks and Branches

Region	Change in Banks (%)	Change in Branches (%)
New England	-20.97	39.31
Mideast	-27.5	19.64
Great Lakes	-32.82	-2.51
Plains	-22.06	25.71
Southeast	-23.46	20.65
Southwest	-22.8	56.83
Rocky Mountain	-32.27	34.67
Far West	-29.78	30.94

SHIFT-SHARE ANALYSIS

The shift-share analysis breaks down changes in regional economic. Shift-share analysis is used to measure the movement or the "shift" of economic activity within a certain region. Any change has three distinct components. These components are national growth, industrial mix, and competitive share components (Selting and Loveridge, 1994).

The national growth component is used to examine the external factors that affect a region's economic trends. The national growth component assumes that the regional economy is identical to the national economy. This is likely the least useful of the three measures as it is based on the nation's average growth rate and does not account for the sector's growth rate (nationally or regionally) nor does it account for the regional growth rate. The nation's employment grew 1.2% from 2001-2012. This figure is about 3.9 percentage points lower than the employment growth for the banking sector over the same time period.

The second component is the industrial mix. This component is used to measure changes in the regional economy. Ordinarily, the industrial mix would sum the employment across all sectors multiplied by the difference in the national growth rate for a particular sector and the whole economy. However, it has been modified since the goal of this paper is to examine only one sector of the economy. This allows for an examination of regional employment changes that come from the concentration of a regional economy in a region growing faster or slower than the national average (Nazara and Hewings, 2004).

The final component of shift-share analysis is competitive share. Competitive share examines whether or not a regional economy can capture an increasing (decreasing) share of growth for a particular industry. This is essentially looking to see if a particular sector in a particular region is growing faster or slower than that industry is at the national level. If a region has a positive competitive share then that sector in that region has a competitive advantage and is capturing more of the market for that industry than other regions (Schaffer, 1976). However, it must be noted that banking services tend to be local and the ability of one region to gain an extreme competitive advantage and take banking services away from another is limited.

For ease of exposition, call 2001-2007 the first period and 2007-2012 the second period. One of the criticisms of shift share is that it is dependent on starting and ending points. We included 2007 to avoid beginning the analysis during a recession. Had we chosen 2008 as a starting point, all regions appear to have grown relative to that point.

From this analysis we can imply that banking employment in each region is doing several things. First of all, in column 1 of Table 3 we see that each region is gaining some banking employment simply as a result of the growth of the nation as a whole in the first period. In the second period every region lost banking employment as a result of national growth factors. This is to be expected given the 2007-2009 recession. Secondly, looking in the competitive growth column, the shift-share results suggest that three of the regions were losing banking employment due to competitive factors in the first period. In the second period, the Southeast and Rocky Mountain regions joined New England, the Mideast, and the Great Lakes in losing banking employment for competitive growth reasons. These competitive factors could be anything from wages in the regions to tax structure and regulation differences from region to region (Bertrand, Schoar, and Thesmar, 2007). Such factors will be captured later in the regression analysis, using regional dummy variables.

Table 3. Results of Regional Shift-Share Analysis

Shift Share 2001-2007			
Region	**National Growth**	**Industrial Mix**	**Competitive Growth**
New England	2,448	1,730	-5,428
Mideast	9,152	6,470	-26,052
Great Lakes	9,521	6,730	-18,012
Plains	5,150	3,641	6,886
Southeast	13,519	9,557	13,301
Southwest	5,932	4,193	16,054
Rocky Mountain	1,720	1,216	6,009
Far West	6,449	4,559	7,242

Shift Share 2007-2012			
Region	**National Growth**	**Industrial Mix**	**Competitive Growth**
New England	-1,647	507	-4,666
Mideast	-5,989	1,843	-3,708
Great Lakes	-6,499	2,000	-6,987
Plains	-4,009	1,233	7,365
Southeast	-10,381	3,194	-7,082
Southwest	-4,858	1,495	14,623
Rocky Mountain	-1,449	446	-2,508
Far West	-4,978	1,532	2,962

Since New England, the Mideast, and the Great Lakes regions all have negative competitive share values, there may be some factors making them less competitive than the Plains, Southeast, Southwest, Rocky Mountain, and Far West regions in terms of banking employment in the first period. The same can be said of New England, the Mideast, Great Lakes, Southeast, and Rocky Mountain regions in the second period.

ESTIMATING SECTORAL EMPLOYMENT

In the previous section we examined a shift-share analysis of banking employment, which although it presents one view of what has happened, it does not estimate a causal link for what has happened to banking employment. To get a better look at the causation of the redistribution in banking employment a regression model is used. The model examines trends in commercial banking employment for all 50 states from 2001-2012 using data from the FDIC and BLS.

$$Emp=\beta_0+\beta_1 lnassets_{i,t}+\beta_2 lnInstitutions_{i,t+}\beta_3 lnBranches_{i,t}+\beta_4 lnUnemp_{i,t+}\beta_5 lnPop_{i,t}+\beta_6 Pop2_{i,t}+\beta_7 Year_{i,t}+\beta_8 Region_{i,t}+\varepsilon_{i,t}$$

Where:

Emp: the natural log of banking employment in state i at time t
Assets: the natural log of bank assets in state i at time t
Branches: the natural log of the number of bank branches in state i at time t
Unemp: the natural log of the unemployment rate of state i at time t
Pop: the natural log of the population of state i at time t
Pop2: the value of the population in state i at time t, squared
Year: Dummy variables for the year of the observation
Region: Regional dummy variables
$\varepsilon_{i,t}$: A composite error term capturing everything not included in the model

The non-discrete variables have been transformed using natural logs to yield coefficients that are partial elasticities rather than levels. This is so that we can interpret the coefficients given by the model as giving us the proportional change in the dependent variable for a given proportional change in the independent variable.

The coefficient on Assets is expected to be positive. A state with more assets would in theory require more bank employees to manage those assets and carry out transactions on the assets. This idea also carries over to branches which, as the number of branches increases, it would be expected that the number of employees would increase as well, and vice versa. This idea may not hold however, as shown in Fung (2006), if certain conditions are met in terms of labor-saving technologies that could decrease banking employment even during periods of consolidation.

The coefficient on banks is less intuitive. It would instinctively make sense for the number of bank employees to decrease as the number of banks would decrease, but this may not always be true. As found by Berger and Humphrey (1992) and Rhoades (1993), the coefficient on banks should be close to zero, so consolidation should have little effect on employment levels.

Table 4. Linear Regression Estimation Results for Commercial Banking Employment

2001-2007				2007-2012			
Number of obs = 350	R-squared = 0.9706			Number of obs = 300	R-squared = 0.9689		
Banking Employment	Coefficient	t		Banking Employment	Coefficient	t	
Assets	0.12	8.38	***	Assets	0.07	7.13	***
Institutions	0.16	7.87	***	Institutions	0.13	5.71	***
Branches	-0.02	-0.38		Branches	0.04	0.58	
Unemployment	0.14	2.54	**	Unemployment	0.26	3.65	***
Population	0.69	13.39	***	Population	0.73	13.16	***
Population^2	1.31E-16	2.47	**	Population^2	1.08E-16	2.1	**
2002	-0.01	-0.31		2008	-0.06	-1.62	
2003	-0.01	-0.41		2009	-0.20	-3.45	***
2004	1.56E-03	0.04		2010	-0.26	-4.29	***
2005	0.01	0.4		2011	-0.25	-4.38	***
2006	0.04	1.14		2012	-0.23	-4.43	***
2007	0.06	1.46		Intentionally left blank			
New England	-0.12	-2.26	**	NewEngland	-0.07	-0.92	
Mideast	0.09	1.75	*	Mideast	0.06	1.24	
Great Lakes	-0.04	-1.42		GreatLakes	-0.03	-0.9	
Plains	0.09	2.34	**	Plains	0.31	8.2	***
Southwest	-0.03	-0.64		Southwest	0.10	2.04	**
Rocky Mountain	-0.06	-1.31		RockyMountain	0.08	2.34	**
FarWest	-0.21	-4.57	***	FarWest	-0.20	-3.82	***
Constant	-3.57	-8.8	***	Constant	-3.75	-7.97	***

* denotes significance between 5-10%.

** denotes significance between 1-5%.

*** denotes significance below 1%.

As noted in Kashian and Monaco (1998), the correlation between banks and branches must not be too high or the model will suffer from multicollinearity. Kashian and Monaco 1998) reported a coefficient of .32 between banks and branches for the first half of the 1990s. Since then, the correlation between the number of banks and branches has risen. For the period 2001-2012, the correlation between the number of banks and branches was .49. Although this is not high enough to create collinearity problems, it is interesting to note the increase in the coefficient from Kashian and Monaco (1998), to our estimated correlation coefficient in the 2001-2012 time period.

RESULTS

The estimated coefficients tend to have the predicted signs in both time periods; however, some of the signs are surprising, especially given the estimations of the shift-share analysis. It was hypothesized that the coefficient on the number of banks would be close to zero, but this is not seen in the data. Since the coefficient on banks is given as an elasticity, we can interpret it as a 1% increase in the number of banks results in a .16% increase in banking employment in the first period. In the second period this coefficient changes to .13, so a 1% increase in banks will result in a .13% increase in banking employment.

The second surprise given is a lack of significance on the coefficient for the number of branches. In previous studies the number of branches has been a fairly significant variable (Kashian, R., & Monaco, K., 1998), however here it does not show significance on banking employment here.

A surprise with respect to the shift-share analysis comes when examining the regional dummy variables. The New England, Mideast, Plains, and FarWest region dummies are all significant in the first epoch. However, the shift share analysis showed a competitive disadvantage for banking employment in the Mideast region for both periods. The results of the regressions show a positive significant coefficient on the Mideast region in the first epoch and a negative significant coefficient on the FarWest region once again in both epochs. These apparently contradictory results may be due to the Southeast being the omitted region in the regression because the region coefficients change depending upon which region is selected for omission to prevent collinearity issues. Alternatively, the regressions include a larger number of variables, so the shift-share analysis may have suffered from omitted variable bias.

One of the possible explanations for the different results in the regression (relative to the shift share analysis) is that the shift share analysis suffers from omitted variable bias such that a the shift share model incorrectly leaves out one or more important causal factors. The 'bias' is created when the model compensates for the missing factor by over- or underestimating the effect of one of the other factors.

Looking back to the shift-share analysis, the Mideast region was estimated to have lost around 26,000 commercial banking jobs during the first time period and about 3,700 in the second time period due to a competitive disadvantage. This led to the hypothesis that the coefficient on the Mideast region would be negative over the time periods. However, when looking at the regression results the coefficient is positive. This leads to the idea that rather than the Mideast having a competitive disadvantage over other regions, the Mideast actually has a competitive advantage. Then why did the Mideast region lose commercial banking jobs? It appears to be due to a rebalancing of the commercial banking industry.

Looking at the change in commercial banking employment in both time periods it is clearly observable that the Mideast lost jobs, while the Plains and FarWest regions gained jobs. The Plains region, like the Mideast, has a positive coefficient which would suggests that they would gain jobs in commercial banking due to some advantage inherent with being in the Plains region. The FarWest on the other hand, has negative coefficients. This leads to the possibility that they would be penalized in commercial banking employment due to some inherent factor associated with the FarWest region (i.e. a competitive disadvantage). However, this disadvantage was not observed when examining commercial banking employment for the FarWest region; indeed, the FarWest region saw an almost 12% increase in their commercial banking employment over the entire timeframe.

These observations lead to the notion that the U.S. economy underwent a rebalancing of commercial banking employment from 2001-2012. This implies that the economy of the Mideast region was oversaturated with commercial banking employment. It also implies that the FarWest region's employment in commercial banking was below the point that it should have been at. Since the Plains region had a positive coefficient in the regression and was predicted to have a competitive advantage by the shift-share model, it cannot be assumed that this regional economy was out of balance. Conversely, it also cannot be said that the gains in commercial banking employment in the Plains region were entirely the result of a competitive advantage, given that a rebalancing appears to have occurred at this time.

CONCLUSION

There are two important findings resulting from this study. The first of these findings is, that the number of banks has a positive relationship with banking employment. Thus, consolidation can in fact negatively affect the number of commercial banking employees.

The second important finding is that from 2001-2012 the U.S. economy may have undergone a rebalancing insofar as commercial banking employment is concerned. The results of the shift-share in conjunction with the OLS regression indicate a discrepancy between where there should be a competitive advantage and where employment shifted to.

REFERENCES

Barff, R. D. (1988). Dynamic shift share analysis. *Growth and Change*, *19*(2), 1-10.

Berger, A., & Humphrey, D. (1992). Measurement and efficiency issues in commercial banking. In Z. Griliches (Ed.), *Output measurement in the service sectors, Chapter 7. National Bureau of Economic Research.*

Bertrand, M., & Schoar, A. (2007). Banking deregulation and industry structure: Evidence from the French banking reforms of 1985. *The Journal of Finance*, *62*(2), 597-628.

Berzeg, K. (1978). The empirical content of shift-share analysis. *Journal of Regional Science*, *18*(3), 463-468.

Blien, U., & Wolf, K. (2002). Regional development of employment in eastern germany. an analysis with an econometric analogue to shift-share techniques. *Papers in Regional Science*, *81*(3), 391-414.

Chilian, M. N. (2012). Evolution of Regional and Sub-Regional Disparities in Romania–A Sectoral Shift-Share Analysis. *Romanian Journal for Economic Forecasting*, (1), 187-204.

Fung, M. (2006). Are labor-saving technologies lowering employment in the banking industry?. *Journal of Banking and Finance*, *30*(1), 179-198.

Kashian, R., & Monaco, K. (1998). Banking employment in the 1990s: Regional effects of consolidation. *Growth and Change*, *29*(2), 145-156.

Mayor, M., Lopez, A., & Perez, R. (2007). Forecasting regional employment with shift–share and arima modelling. *Regional Studies*, *41*(4), 543-551.

Nazara, S., & Hewings, G. (2004). Spatial structure and taxonomy of decomposition in shift-share analysis. *Growth and Change*, *35*(4), 476-490.

Rhoades, S. (1993). Efficiency effects of horizontal (in-market) bank mergers. *Journal of Banking and Finance*, *17*(2-3), 411-422.

Schaffer, R. (1976). *Changes in Wisconsin's nonfarm private employment, 1962-1972 : a shift and share analysis*. Informally published manuscript, College of Agricultural and Life Sciences, University of Wisconsin-Madison, Madison, WI.

Selting, A., & Loveridge, S. (1994). Testing dynamic shift share. *Journal of Regional Analysis and Policy*, *24*(1), 23-41.

Stilwell, F. (1969). Regional growth and structural adaptation. *Urban Studies*, *6*(2), 162-178.

In: Banking
Editor: Janice P. Simmons

ISBN: 978-1-62948-897-4

Chapter 3

WHO DEVELOPS POSTTRAUMATIC STRESS SYMPTOMS FOLLOWING A BANK ROBBERY? A NATIONAL COHORT STUDY

Maj Hansen** *and Ask Elklit
National Centre of Psychotraumatology, Institute of Psychology, University of Southern Denmark, Odense, Denmark

ABSTRACT

Each year, numerous bank robberies take place worldwide. Even so, only few studies have investigated the psychological sequelae of bank robbery and little is known about the risk factors associated with the development of posttraumatic stress disorder (PTSD) following this potentially traumatic event. Knowledge about risk factors related to PTSD may allow for preventive measures to be taken against the development of PTSD and reduce the large cost associated with the disorder. Furthermore, the few existing studies are characterized by several limitations such as the use of small convenience or self-selecting samples. To overcome these limitations, we investigated the estimated prevalence rate of acute stress disorder (ASD) and PTSD and predictors of PTSD severity in a Danish cohort study of all bank employees exposed to robbery during one year (N = 614). A total of 450 employees (73 %) filled out the first questionnaire a week after the robbery (T1). Of these, 371 employees (82 %) filled out the second questionnaire six months

* Maj Hansen: E-mail: mhansen@health.sdu.dk.

after the robbery (T2). Results showed that 11.1% of the participants suffered from ASD (T1) and 6.2 % suffered from PTSD (T2). The results of a hierarchical regression analysis showed that 51 % of the variance in PTSD severity could be explained with only peritraumatic dissociation, ASD severity, and negative cognitions about self being significant predictors. However, interpretation of the role of peritraumatic dissociation in the regression model was hampered by a statistical artifact (negative suppression). The strongest predictors of PTSD severity were ASD severity followed by negative cognitions about self. This indicates that victims with high levels of ASD are at increased risk of developing PTSD following bank robbery. Thus, screening for PTSD following bank robbery should focus on ASD severity. In accordance with a cognitive model of PTSD and the DSM-5, the results show that posttraumatic negative cognitions about self also seem to play a pivotal role in the development of PTSD following bank robbery. This indicates that it may be possible to develop preventive cognitive interventions targeted at bank employees that focus on changing negative cognitions about self.

Each year, numerous bank robberies take place worldwide (the Danish Bankers Association, 2013; Gilioli, Campanini, Fichera, Punzi, & Gassitto, 2006). The exact number of annual Danish bank robberies have varied the past 15 years (69-222) and fortunately the number is currently low (71 in 2012; the Danish Bankers Association, 2013). However, there are numerous victims per robbery as victims can be both employees, customers, random passing people, security guards, police officers or other types of rescue personal. This chapter focuses solely on the psychological impact of bank robbery on the employees, who are often the primary victims of the bank robbery. During a bank robbery the employees are often threatened at gun point to hand over the cash. Most people report feelings of life threat, helplessness and intense horror, when they are held at gun point (Elklit, 2002). Even so, only few studies have investigated the psychological sequelae of bank robbery and little is known about the risk factors associated with the development of posttraumatic stress disorder (PTSD) following this potentially traumatic event (Hansen & Elklit; 2011; 2013; Kamphuis & Emmelkamp, 1998; Ladwig et al., 2002; Miller-Burke, Attridge, & Fass, 1999). The limited research shows that bank robberies are potentially traumatizing events associated with both immediate and long-term posttraumatic symptoms. However, the existing bank robbery studies have all been based on convenience samples which have identified participants through crisis aid services or a particular bank service. Furthermore, most of the samples have been relatively small (N= 53-152).

Indeed, only one study can be regarded as having a large sample (N = 303; Kamphuis & Emmelkamp, 1998). This creates problems with the validity and reliability of the results as it is difficult to ensure that the results are representative and generalizable when they are based on rather small and self-selected samples. Additionally, only one study has investigated the prevalence rates of acute stress disorder (ASD) and PTSD as specified by the American Diagnostic and Statistical Manual of Mental Disorders (DSM-IV-TR; APA, 2000). We previously estimated that 14.5 % of bank employees suffered from an acute stress disorder (ASD) diagnosis and that 6.8 % of employees suffered from a PTSD diagnosis in an earlier convenience study of bank robbery victims (Hansen & Elklit, 2011; 2013). Studies of other forms of robbery and nonsexual assault have found higher prevalence rates of both ASD (17-25%) and PTSD (20-30%; Brewin, Andrews, Rose, & Kirk, 1999; Elklit, 1999; 2002; Elklit & Brink, 2004; Kleim, Ehlers, & Clucksman, 2007).

The DSM-IV-TR (APA, 2000) defines ASD and PTSD as diagnoses that follows experiencing, witnessing, or being confronted with an event that involves death, physical injury, or a threat to the physical integrity of self or others (criterion A1), in which the individual reacts with feelings of fear, helplessness, or intense horror (criterion A2). The diagnostic criteria for both diagnoses state a requirement of symptoms of intrusion/re-experiencing, avoidance, and arousal. Symptoms of intrusion/re-experiencing concern that the traumatic event is somehow re-experienced for instance through thoughts, dreams, or flashback episodes. The avoidance criterion concerns efforts to avoid stimuli associated with the traumatic event for instance thoughts, feelings, activities, places or people associated with the traumatic event. Symptoms of arousal concern for instance sleep difficulties, hypervigilance, poor concentration, and irritability. Also, both diagnoses require functional impairment. The main differences between the diagnostic criteria for PTSD and ASD are the duration of the symptoms and the dissociation symptom cluster in the ASD diagnosis (see the later section on peritraumatic dissociation for definition). ASD describes acute posttraumatic symptoms (two days to one month) following traumatic exposure, whereas PTSD describes long-term posttraumatic symptoms (more than a month) following traumatic exposure. The ASD diagnosis requires that the individual has at least three of five described dissociative symptoms, which the PTSD diagnosis does not. Beyond the time criterion and the dissociation criterion, several additional differences need to be mentioned. As pointed out by Bryant and Harvey (1997) only the PTSD diagnosis requires that intrusion (i.e. re-experiencing in the ASD diagnosis) involves or cause distress. At the same time, the PTSD diagnosis requires the presence of at least three symptoms of avoidance and two symptoms of arousal, whereas the ASD diagnosis only states a diffuse requirement of marked arousal and avoidance. The costs of PTSD as well as ASD are not only related to the individual, but the workplace, family and society are also affected. Thus, PTSD is associated with a large range of costs such as for instance costs associated with reduced quality of life, functional incapacities, sick leave, work termination, reduced productivity, and increased need of health care and social support (Gilioli et al., 2006).

Despite exposure to bank robberies being a potential traumatic event associated with the development of posttraumatic symptoms, only a few studies have investigated risk factors for posttraumatic symptoms in order to identify victims at risk of developing PTSD (Hansen & Elklit, 2011; 2013; Kamphuis & Emmelkamp, 1998; Ladwig et al., 2002; Miller-Burke et al., 1999). Knowledge about risk factors for the development of PTSD may facilitate early treatment and preventive interventions, which may hinder or minimize the development of PTSD and its associated costs (Litz & Maguen, 2007). There is a plethora of research investigating risk factors associated with posttraumatic symptomatology after a long range of other forms of traumatic exposure, which has resulted in the identification of several risk factors for posttraumatic symptomatology (Brewin, Andrews, & Valentine, 2000; Ozer, Best, Lipsey, & Weiss, 2003). These factors can be grouped temporally according to the time of the traumatic exposure as pre-trauma, peri-trauma and post-trauma factors (Ladwig et al., 2002). Pre-trauma factors are factors present before the traumatic exposure, peri-trauma factors are factors appearing during the traumatic exposure, and post-trauma factors are factors, which present themselves after the traumatic exposure. Furthermore, research indicates that multiple factors rather than single factors may play a part in determining the psychological impact of traumatic exposure; including bank robbery (Brewin et al., 2000; Hansen & Elklit, 2011; 2013; Kamphuis & Emmelkamp, 1998; Miller-Burke et al., 1999; Ozer et al., 2003). At the same time, two large meta-analyses of risk factors for posttraumatic symptomatology indicate that there is variation in risk factors across trauma types (Brewin et al., 2000; Ozer, et al., 2003). Thus, the identified risk factors for posttraumatic symptomatology following different forms of traumatic exposure may not constitute risk factors for posttraumatic symptoms following bank robbery and vice versa.

To summarize, there is a great need of studies seeking to clarify the psychological impact of bank robbery and risk factors associated with PTSD following bank robbery. In particularly, we need to know how many employees suffer from ASD and PTSD and how we can identify those at risk of developing PTSD in order to facilitate early treatment and reduce the suffering and large costs associated with PTSD. To overcome the limitations of the previous research in bank robberies, we investigate the psychological impact of bank robbery one week (ASD) and six months (PTSD) after the robbery and predictors of PTSD severity in a national cohort study of bank employees exposed to bank robbery. In the following next three sections "pre-trauma risk factors", "peri-trauma risk factors", and post-trauma risk factors", this chapter resembles a review of research concerning the predictors of PTSD selected for the current study. Afterwards, this chapter takes more form of an empirical study investigating the prevalence rates of ASD and PTSD following bank robbery and the prediction of PTSD severity.

PRE-TRAUMA RISK FACTORS

Research on the psychological impact of bank robberies and other forms of robberies at work has shown discrepancies regarding whether or not the pre-trauma variables; younger age, female sex, and prior traumatic exposure are significant risk factors for posttraumatic symptomatology (Elklit, 2002; Hansen & Elklit, 2011; 2013; Ladwig et al., 2002; Miller-Burke et al., 1999). This is in accordance with the Brewin et al. (2000) meta-analysis of risk factors for PTSD showing that age, sex, and prior traumatic exposure are generally significant risk factors for PTSD but only with small effect sizes. However, age does not seem to emerge as a significant risk factor associated with posttraumatic symptomatology following bank robbery (Hansen & Elklit, 2011; 2013; Ladwig et al., 2002; Miller-Burke et al., 1999). At the same time, female sex is found to be associated with increased posttraumatic symptoms in some robbery studies (Elklit, 2002; Hansen & Elklit, 2011; 2013) but not in others (Miller-Burke et al., 1999). Similarly, there are discrepancies in robbery studies regarding whether different forms of prior traumatic exposure are significantly associated with increased posttraumatic symptoms or not (Hansen & Elklit, 2011; 2013; Miller-Burke et al., 1999). The discrepancies may be due to differences in which types of prior traumatic exposure that are measured, as all trauma types are not necessarily equal in their effect (Ozer et al., 2003). Furthermore, according to Brewin et al. (2000) it is possible that the effect of these pre-trauma risk factors on the development of PTSD may be distal rather than proximal. Thus, the effect of these pre-trauma factors may have an indirect effect on the development of posttraumatic symptoms through later factors associated with the trauma. In addition, recent exposure to major life events has been found associated with increased posttraumatic symptomatology after robberies (Elklit, 2002; Hansen & Elklit, 2011; Kamphuis & Emmelkamp, 1998).

According to cognitive models of responses to traumatic exposure, the individual's appraisal of traumatic exposure and their capacity to respond to that experience is essential to how they will adapt to the trauma (Ehlers & Clark, 2000). Research indicates that these cognitive responses not only include posttraumatic cognitions (see later section on posttraumatic cognitions) but also maladaptive thinking associated with panic (cf., Nixon & Bryant, 2005). Indeed, anxiety sensitivity has been found highly associated with posttraumatic symptoms (Bryant & Panasetis, 2001; Taylor, Koch, & McNally, 1992). Thus, the current study also investigates the association between proneness to panic reactions (anxiety sensitivity, i.e., a heighted awareness of bodily sensations and the tendency to interpret these sensations catastrophically) and posttraumatic symptoms. This association has not previously been investigated following bank robbery.

Furthermore, neuroticism has also been found to be significantly and positively associated with posttraumatic symptomatology following a wide range of traumatic exposure (Alexander & Wells, 1991; Breslau & Schultz, 2013; Christiansen & Elklit, 2008; Parlow, Jorm, & Christensen, 2006). Indeed, neuroticism is said to measure vulnerability to the development of psychiatric disorders (Breslau & Schultz, 2013). Neuroticism is defined by Costa and McCrae (1987 p. 301) as "a broad dimension of individual differences in the tendency to experience negative, distressing emotions and to possess associated behavioral and cognitive traits". The neuroticism dimension is defined by personality traits such as fearfulness, social anxiety, low self-esteem, irritability, poor inhibition of impulses, and helplessness. Neuroticism has previously been investigated in relation to bank robbery but not as a risk factor for posttraumatic symptomatology (Kamphuis & Emmelkamp, 1998).

PERI-TRAUMA RISK FACTORS

Several different aspects of trauma severity have been found associated with the development of posttraumatic symptoms after bank robbery (Hansen & Elklit, 2011; 2013; Ladwig et al., 2002; Miller-Burke et al., 1999) and other forms of traumatic exposure (Brewin et al., 2000; Ozer et al., 2003). Perceived life threat during the trauma is one aspect of trauma severity which has been found associated with the development of posttraumatic symptoms after bank robberies as well as other forms of traumatic exposure (Hansen & Elklit, 2011; 2013; Kleim et al., 2007; Ladwig et al., 2002; Miller –Burke et al., 1999; Ozer et al., 2003). Proximity to the robbery has also been associated with the development of posttraumatic symptoms following bank robbery (Hansen & Elklit, 2011; 2013; Miller-Burke et al., 1999). Furthermore, perceived helplessness, fear or horror, and/or fear of dying (the DSM-IV-TR A2 criterion) have also been found to be associated with the development of posttraumatic symptoms following bank robbery and other forms of nonsexual assault (cf. Brewin, Andrews & Rose, 2000; Hansen & Elklit, 2011; 2013). However, recently the utility of the A2 criterion has been questioned as it seems to have limited capacity to predict PTSD following traumatic exposure (Friedman, Resick, Bryant & Brewin, 2010). Thus, in the newly released DSM-5 the A2 criterion has been removed from both the ASD and the PTSD diagnoses (APA, 2013). However, the A2 criterion may still have utility in relation to predicting PTSD following bank robbery as there may be variation in risk factors across trauma types.

Research is beginning to show increasingly promising results for tonic immobility, peritraumatic panic, and peritraumatic dissociation as risk factors for developing PTSD following different forms of traumatic exposure (Bryant, Brooks, Silove, Creamer, O'Donnell, & McFarlane, 2011; Marx, Forsyth, Gallup, Fusé, & Lexington, 2008; Ozer et al., 2003). Tonic immobility is conceptualized as an evolutionary survival strategy and is considered to be the last defense in a series of defense reflexes (i.e. freeze, flight, fight, and tonic immobility) against entrapment by a predator, when it is no longer possible to escape or resist (Gallup, 1977; Marx et al., 2008). More specifically, tonic immobility is characterized by involuntary immobility, analgesia, and unresponsiveness to external simulation. Recent research indicates that tonic immobility occurs in people in mortal danger and under physical restraint (or perceived immobility to escape; Heidt, Marx, & Forsyth, 2005; Kunst, Winkel & Bogaerts, 2011; Marx et al., 2008; Rocha-Rego et al., 2009; Volchan et al., 2011). Furthermore, tonic immobility has been found associated with increased PTSD symptomatology after different forms of traumatic exposure including nonsexual assault (Heidt et al., 2005; Kunst et al., 2011; Marx et al., 2008; Rocha-Rego et al., 2009; Volchan et al., 2011). Several possible explanations for the relationship between tonic immobility and the development of PTSD have been put forward for instance that tonic immobility and PTSD are the by-products of the same physiological changes or that tonic immobility hinders more adaptive defensive responses during and after the traumatic exposure (Marx et al., 2008). However, it is unclear which precise mechanism lies behind the relationship between tonic immobility and the development of PTSD (Marx et al., 2008). At the same time, tonic immobility has not previously been investigated following bank robbery.

The occurrence of panic during traumatic exposure (i.e., peritraumatic panic) is very common (Bryant et al., 2011). Shortness of breath, choking, sweating, nausea, chest pain, dizziness, and fear of losing control are examples of symptoms of panic. Research indicates that more than half of all trauma survivors report at least four symptoms of panic during a traumatic exposure, which is the required number to meet the diagnostic criteria for a panic attack according to the DSM-IV (APA, 1994; Bryant & Panasetis, 2001). Peritraumatic panic may be related to the development of PTSD, as panic may result in a more distressing experience or may condition trauma related cues that can trigger subsequent anxiety (Bryant & Panasetis, 2001; Falsetti, Resnick, Dansky, Lydiard, & Kilpatrick, 1995). Indeed, peritraumatic panic has been found associated with the development of posttraumatic symptoms following different forms of traumatic exposure (Bryant et al., 2011; Bryant & Panasetis, 2001; Nixon & Bryant, 2003; Rocha-Rego et al., 2009). However, the relationship between posttraumatic symptoms and panic has not previously been investigated following bank robbery.

Peritraumatic dissociation refers to changes in consciousness during traumatic exposure. In the DSM-IV peritraumatic dissociation is defined as a subjective feeling of emotional numbness, detachment from others, reduced responsiveness to one's surroundings, depersonalization, and derealization during the traumatic exposure (APA, 1994). Peritraumatic dissociation is assumed to be associated with PTSD, because it prevents adequate processing and thus adequate integration of the traumatic recollections (Breh & Seidler, 2007). Indeed, peritraumatic dissociation is common during traumatic exposure and has been found a risk factor of PTSD following a wide range of traumatic exposure (Breh & Seidler, 2007; Bryant, 2009; Ozer et al., 2003) including one study of bank robbery (Ladwig et al., 2002). Indeed, Ozer et al.'s (2003) meta-analysis found that peritraumatic dissociation emerged as the strongest risk factor (weighted overall effect size $r = .35$) out of seven investigated risk factors. Furthermore, Breh and Seidler's meta-analysis (2007) of prospective studies of peritraumatic dissociation and PTSD largely replicated the results of the Ozer et al. (2003) meta-analysis (weighted overall effect size $r = .34$). However, a recent review of methodologically sound prospective studies of the predictive value of peritraumatic dissociation indicates that peritraumatic dissociation is not an independent predictor of

PTSD (van der Velden & Wittmann, 2008). The relationship between peritraumatic dissociation and PTSD becomes non-significant when controlling for other factors such as ASD (van der Velden & Wittmann, 2008). The relationship between peritraumatic dissociation and posttraumatic stress symptoms following bank robbery has only previously been investigated in relation to the ICD-10 acute stress reaction and PTSD (WHO, 1992) but not in relation to ASD and PTSD as specified by the DSM-IV-TR.

Post-Trauma Risk Factors

The Acute Stress Disorder (ASD) diagnosis was introduced into the DSM-IV with a twofold purpose. The first purpose was to recognize posttraumatic stress occurring within the first month after traumatic exposure and the second purpose was to identify victims at risk of developing PTSD (APA, 1994). Several studies have investigated the predictive power of different aspects of the ASD diagnosis on PTSD (Bryant, 2011). According to Bryant's (2011) systematic review, the ASD diagnosis has limited capacity to predict PTSD. However, Bryant's (2011) review only included two studies of nonsexual assault (Brewin et al., 1999; Elklit & Brink, 2004) that both indicated that ASD predicts PTSD better following nonsexual assault than other forms of traumatic exposure.

At the same time, we previously found that ASD severity rather than the ASD diagnosis seems to be a strong predictor of PTSD severity following bank robbery (Hansen & Elklit, 2013). The strong relationship between ASD severity and PTSD severity was supported by ASD severity being the strongest predictor of PTSD severity (explaining 40 % of the variance in PTSD severity). ASD severity therefore appears to be a strong risk factor for posttraumatic symptoms following bank robbery.

As mentioned the Ehlers and Clark's (2000) cognitive model of trauma response state the individual's appraisal of the traumatic exposure and their capacity to respond to it is essential to how they will adapt to the trauma. Catastrophic or negative perceptions about the trauma, its implications, and the individual's own capacity to handle the trauma can lead to a sense of current threat which can contribute to the development of PTSD. The negative perception prevents the individual from processing corrective information, which could help the individual recognize that the threat has passed. Indeed, negative cognitive responses to traumatic exposure have been found important in the prediction of posttraumatic symptoms following different types of traumatic exposure (Bryant, 2003; Karl, Rabe, Zöllner, Maercker, & Stopa, 2009; Nixon & Bryant; 2005). Low perceived safety (a form of cognition) has also been found predictive of both ASD and PTSD following bank robbery (Hansen & Elklit, 2011; 2013).

Several studies indicate that cognitive models are better predictors of PTSD than either ASD or empirically derived variables such as peritraumatic dissociation (Karl et al., 2009; Kleim et al., 2007). Furthermore, as mentioned previously research also indicates that these cognitive responses concern maladaptive thinking associated with panic. Thus, we investigate negative cognitions about self, megative cognitions about the world, self-blame, and negative cognitions about bodily sensations as risk factors of PTSD. These negative cognitions have not previously been investigated following bank robbery.

Social support has been found associated with the development of posttraumatic symptoms after a wide range of traumatic exposure; including bank robbery (cf. Brewin et al., 2000; Miller-Burke et al., 1999; Ozer et al., 2003; Yap & Devilly, 2004). Social support can be defined in many ways, but the concept usually refers to positive aspects of social interaction (Holeva, Tarrier & Wells, 2001). Research indicates that whether or not social support constitutes a protective factor for posttraumatic symptoms depends on the quality of and the satisfaction with the social support rather than the simply the availability of social support (Andrews, Brewin & Rose, 2003; Elklit & Brink, 2004; Miller-Burke et al., 1999). Receiving social support can be a burden on the victims if it is ambiguous, criticizing or, guilt invoking support (Buunk & Hoorens, 1992; Yap & Devilly, 2004). Robbery studies have also found that positive aspects of social support are protective against the development of posttraumatic symptoms, and that negative aspects of social support are risk factors for posttraumatic symptoms (Elklit, 2002; Hansen & Elklit, 2011; 2013; Miller-Burke et. al., 1999; Richards, 2000). Thus, we investigate perceived positive social support, perceived negative social support, and social support satisfaction.

Finally, decades of research have established the importance of coping when dealing with a stressful or traumatic event. Indeed, research has found coping associated with the development of posttraumatic symptoms following a wide range of traumatic exposure including bank robbery (Agaibi & Wilson, 2005; Bödvarsdóttir & Elklit, 2004; Kamphuis & Emmelkamp, 1998; Schnider, Elhai, & Gray, 2007). Coping is often defined according to Lazarus and Folkmann's (1984) theory on stress as cognitive and behavioral efforts to manage situational demands that are appraised as exceeding the resources of the individual (Monat & Lazarus, 1991). Many different coping styles have been identified, but they can often be grouped as either positive (e.g. rational or problem-focused coping) or negative coping (e.g. avoidant or emotion-focused coping) as positive coping styles are usually found protective against the development of PTSD, whereas negative coping styles are found to constitute risk factors for PTSD (Bödvarsdóttir & Elklit, 2004; Kamphuis & Emmelkamp, 1998; Schnider et al., 2007).

THE CURRENT STUDY

The aim of the current study is two-fold. The first aim is to investigate the psychological impact of bank robbery by estimating the ASD and PTSD prevalence rates. The second aim is to investigate the above stated predictors for PTSD severity following bank robbery. This combination of predictors has not previously been investigated following bank robbery or any other type of traumatic exposure. Due to the extant literature we hypothesize:

- The prevalence rates of ASD and PTSD are consistent with our previous bank robbery study and lower than the prevalence rates found following other forms of non-sexual assault.
- The pre-trauma variables (younger age, female sex, prior traumatic exposure, prior robberies, major life events, anxiety sensitivity, and neuroticism) are all associated with increased PTSD severity.
- The peri-trauma variables (proximity to the robbery, perceived life threat, perceived helplessness, intense horror, tonic immobility, peritraumatic panic, and peritraumatic dissociation are all associated with increased PTSD severity.
- ASD severity, negative cognitions about self, negative cognitions about the world, self-blame, perceived negative social support, avoidant and emotion-focused coping are associated with increased PTSD severity, whereas perceived positive social support, social support satisfaction, and rational coping are associated with decreased PTSD severity.
- Furthermore, based on previous research, we expect that ASD severity and negative cognitions are particular strong predictors of PTSD severity.

METHOD

Procedure and Participants

The current study is part of a large national study of the psychological impact of bank robberies committed in Denmark from April 2010 to April 2011. The study was conducted in collaboration between the Danish Bankers Association, the National Bank of Denmark, all Danish Banks, and the University of Southern Denmark. During this period, 614 bank employees were exposed to robbery. A total of 450 employees (73 %) filled out the first questionnaire a week after the robbery (T1, M = 9.89 days, SD = 6.30).Of these, 371 employees (82 %) filled out the second questionnaire six months after the robbery (T2, M = 191.7, SD = 13.15).The majority of the participants (n = 297, 80.1 %) were in the room during the robbery (i.e. directly exposed), and the remaining participants were in adjoining rooms (i.e. indirectly exposed, n = 74, 19.9 %). As argued by Hansen and Elklit (2011), the stressor criterion A1 of the DSM-IV-TR diagnostic criteria for ASD and PTSD is met during a bank robbery regardless of the presence of the participants, because a bank robbery involves actual or threatened death or serious injury or a threat to the physical integrity of self or others (i.e. colleagues). The majority of the participants (96 %, n = 356) received psychological debriefing, which was conducted in a non-manualized manner by different psychologists and not as a part of the current study. There was no significant difference between the debriefed and non-debriefed group in relation to ASD severity or PTSD severity (measured by the Acute Stress Disorder Scale, Bryant, Moulds & Guthrie, 2000, and the Harvard Trauma Questionnaire, Mollica et al., 1992, $t(369) = .57$, $p = .571$ and $t(369) = .16$, $p = .874$, respectively). Thus, the participants were treated as a homogeneous group. The participants were informed of the purpose of the survey both orally and in writing and instructed to fill out the questionnaires in relation to the index robbery. Participation was voluntary and all necessary permissions for conducting questionnaire surveys, according to Danish Law, were obtained. Furthermore, the study was approved by the PhD review board appointed by the University of Southern Denmark.

Measures

The questionnaire assessed demographic factors and the following variables related to the current study. All of the following described predictors were assessed at T1 and PTSD severity was assessed at T2. Proximity to the robbery was assessed by asking the participants if they had been present in the room during the robbery with answers stated dichotomously (*yes* or *no*). The experience of perceived helplessness, fear of dying, and perceived life threat during the robbery were assessed dichotomously (*yes* or *no*). Prior traumatic exposure was assessed by asking the participants whether they had experienced any of 14 different kinds of traumas applied in the U.S. National Comorbidity Survey (Kessler, Sonnega, Bromet, Hughes, & Nelson, 1995) with a separate scoring of prior bank robberies. Major life changes were assessed in open-ended questions asking the participants if they had experienced any during the past 12 months. The number of experienced changes was summated.

The Danish version of the Anxiety Sensitivity Index (ASI; Reiss, Peterson, Gursky, & McNally, 1986) was used to assess proneness to panic reactions (i.e. negative cognitions about bodily sensations). The ASI is a 16-item self-report questionnaire that measures the participants' belief about the harmfulness of anxiety symptoms on a 5-point Likert type scale (1= *very little* to 5= *very much*). Good test–retest reliability (.75) and construct validity have been reported (Reiss et al., 1986). The reliability coefficient in the current study was satisfactory (Cronbach's α =.89).

Neuroticism was assessed using the Danish version of the neuroticism subscale of the NEO Personality Inventory-revised (NEO PI-R; Costa & McCrae, 2004) short version. The NEO PI-R neuroticism subscale comprises 12 items assessed on a 5-point Likert type scale (0 = *strongly disagree* to 4= *strongly agree*). The NEO PI-R has shown to have good psychometric properties including good reliability (cf. Costa & McCrae, 2004; Rossier, Meyer de Stadelhofen, & Berthoud, 2004) for the neuroticism subscale. The reliability coefficient in the current study was satisfactory (Cronbach's α =.87).

Tonic immobility was assessed using a Danish version of the Tonic Immobility Scale (TIS; Heidt et al., 2005). The TIS originally comprised 10 items assessed on a 7-point Likert Scale type (a low score indicates no tonic immobility and a high score indicates high tonic immobility). However, some of these items assess dissociative and panic-like responses. In order to avoid item overlap between the Physical Reaction Scale, the Peritraumatic Dissociation Questionnaire, and the TIS as the goal is to compare the different predictors of PTSD severity, we only assessed the motor aspects of tonic immobility. Other studies have used the same procedure (Kunst et al., 2011; Rocha-Rego et al., 2009). The four items were: "*froze or felt paralyzed during the event, "unable to move even though not restrained during the event", "unable to call out or scream during the event", and "unable to escape during the event".* Two independent forward-translations into Danish were carried out using the original English version of the TIS. The two versions were carefully reviewed and edited into a final Danish version. The Danish version of the TIS was then back-translated by a different translator and compared with the original English version. Previous studies report good internal consistency Cronbach's α = .88 (cf. Kunst et al., 2011).The reliability coefficient in the current study was satisfactory (Cronbach's α = .85).

Peritraumatic panic was assessed using a modified Danish version of the Physical Reactions Scale (PRS, Falsetti & Resnick, 1992). The PRS is a 16-item self-report questionnaire that measures how severely the participants experienced DSM-IV panic symptoms during the robbery on a 6-point Likert type scale (*0= not at all* to 5= *extremely*). Two independent forward-translations into Danish were carried out using the original English version of the PRS. The two versions were carefully reviewed and edited into a final Danish version. The Danish version of the PRS was then back-translated by a different translator and compared with the original English version. There are currently no psychometric data on the PRS. The reliability coefficient in the current study was satisfactory (Cronbach's α = .91).

Peritraumatic dissociation was assessed using a Danish version of the Peritraumatic Dissociative Experiences Questionnaire (PDEQ; Marmar, Weiss, & Metzler, 1997). The PDEQ comprises 10 items reflecting dissociative responses rated on a 5-point Likert type scale (*0= not at all* to 5= *extremely true*) during the robbery. Two independent forward-translations into Danish were carried out using the original English version of the PDEQ. The two versions were carefully reviewed and edited into a final Danish version. The Danish version of the PDEQ was then back-translated by a different translator and compared with the original English version. The PDEQ has shown satisfactory internal consistency, test-retest reliability (.72), and string convergent reliability (cf. Birmes et al., 2005; Kunst et al., 2011). The reliability coefficient in the current study was satisfactory (Cronbach's α = .87).

ASD severity was assessed using the Danish version of the Acute Stress Disorder Scale (ASDS; Bryant et al., 2000).The ASDS is a 19 item self-report scale with four subscales assessing the four separate symptom clusters dissociation, re-experiencing, avoidance, and arousal, as specified by the DSM-IV-TR. Answers are rated on a 5-point Likert type scale (1 = *not at all* to 5 = *very much*). The ASD symptom clusters were met if the participants endorsed at least one re-experiencing symptom, one avoidance symptom, and one arousal symptom in addition to at least three dissociative symptoms, all indicated by item scores $\geq$ 3 on the ASDS. Previous studies using the Danish ASDS have used this procedure and have reported good reliability, with reliability coefficients of .85 -.96 (cf. Armour, Elklit, & Shevlin, 2011; Elklit & Christiansen, 2010; Hansen & Elklit, 2011) for the total scale. The reliability coefficient in the current study was satisfactory (Cronbach's α = .93).

Posttraumatic cognitions were assessed using a Danish version of the Posttraumatic Cognition Inventory (PTCI; Foa, Ehlers, Clark, Tolin, & Orsillo, 1999). The PTCI is a 33 item self-report scale with three subscales assessing negative cognitions about self, negative cognitions about the world, and self-blame rated on a 7-point Likert Scale (1 = *totally disagree* to 7 = *totally agree*). Scale scores are formed for the three subscales. Two independent forward-translations into Danish were carried out using the original English version of the PTCI. The two versions were carefully reviewed and edited into a final Danish version. The Danish version of the PTCI was then back-translated by a different translator and compared with the original English version. The PTCI has demonstrated high convergent validity with other trauma-related cognition scales and good sensitivity and specificity in relation to correctly identifying PTSD diagnostic status (cf. Beck, Coffey, Palyo, Gudmundsdottir, Miller, & Colder., 2004; Foa et al., 1999). The reliability coefficients in the current study were satisfactory on all subscales (Cronbach's α = .70-.92). Perceived social support after the robbery was assessed using the Danish version of the Crisis Support Scale (CSS; Joseph, Andrews, Williams, & Yule, 1992) which is comprised of seven items. In accordance with previous studies (Andrews et al., 2003), the summated score of the first five items is used as a measure of positive support, item six is used as a single measure of negative support (feeling let down), while item seven is used as a single measure of overall satisfaction with received social support. The answers are rated on a 7-point Likert-type scale (1= *never* to 7= *always*). The Danish version of the CSS has been shown to have good reliability and validity (Elklit, Pedersen, & Jind, 2001). The reliability coefficient on the perceived positive support was unsatisfactory in the current study (Cronbach's α = .59). However, the mean inter-item correlation coefficient (.27) was within the acceptable range (Briggs & Cheek, 1986). Coping style was assessed in relation to how the participants generally deal with stressful events using a revised Danish version of the Coping Styles Questionnaire (CSQ; Roger, Jarvis, & Narajian, 1993). The revised CSQ has 37 items and preserved the four primary coping components: rational (problem-focused) coping, emotion-focused coping, avoidant coping, and detached coping rated on a 4-point Likert scale (1= *never* to 4= *always*; Elklit, 1996). Concordance between the four subscales in the original and revised version is high (67-84%). The four subscales have shown good test-retest reliability (.74-.85; Elklit, 1996). The reliability coefficients in the current study were satisfactory for all subscales (rational Cronbach's α = .79, emotion-focused Cronbach's α = .84, avoidant Cronbach's α = .72,) except for the detached coping subscale (Cronbach's α = .58, interitem correlation mean = .19), which was subsequently excluded from further analysis.

PTSD severity was assessed using the Danish version of the Harvard Trauma Questionnaire part IV (HTQ; Mollica et al., 1992). The HTQ is a 17 item self-report scale with three subscales assessing the three separate symptom clusters: intrusion, arousal, and avoidance as specified by the DSM-IV-TR rated on a four-point Likert scale (1= *not at all* to 4= *all the time).* The PTSD symptom clusters were met if the participants endorsed at least one intrusion symptom, three avoidance symptoms, and two arousal symptoms, all indicated by item scores $\geq$ 3. The Danish version of the HTQ has been used in a wide range of Danish trauma populations with reports of good reliability and validity (cf. Bach, 2003). The reliability coefficient in the current study was satisfactory (Cronbach's α = .92).

Data Analysis

Prior to data analysis the data was screened for errors. The percentage of missing values was small (.00-3.2 %). Thus, the missing data was imputed with the Expectation Maximization (EM) algorithm. The EM algorithm is a demonstrated effective method of dealing with missing data (Bunting, Adamson, & Mulhall, 2002). Preliminary analyses were performed prior to testing the model for predicting PTSD severity. The mutual relationships of the predictor variables (the independent factors) and PTSD severity (the dependent factor; the HTQ total score) were investigated in correlation analyses. The distribution of HTQ scores was a close approximation to a normal distribution and thus did not give cause for concern in conducting the regression analysis. Furthermore, there were no problems with multicollinearity (all tolerance values > .10; all VIF values < 10). Subsequently, a hierarchical multiple regression analysis was conducted to assess the predictive value of the independent variables on the HTQ total score.

RESULTS

Descriptives

The sample descriptives of all the predictor variables and PTSD severity are shown in table 1. The ASD prevalence rate was 11.1 % (*n* = 41), and the PTSD prevalence rate was 6.2 % (*n* = 23).We define subclinical ASD as the presence of three out of the four symptom clusters and subclinical PTSD as the presence of two out of the three symptom clusters as specified by the DSM-IV-TR. The subclinical ASD prevalence rates ranged from 11.1 % to 26.7 % (ASD without dissociation = 26.7 % (*n* = 99), ASD without avoidance = 14.0 % (*n* =52), ASD without re-experiencing = 11.6 % (*n* = 43), and ASD without arousal = 11.1 % (*n* =41)). The subclinical PTSD prevalence rates ranged from 6.2 % to 18.1 % (PTSD without avoidance = 18.1 % (*n* = 67), PTSD without intrusion = 6.7 % (*n* = 25), and PTSD without arousal = 6.2 % (*n* = 23)).

A total of 83 participants (22.4 %) had not been exposed to any prior traumas, whereas 129 participants (34.8 %) had been exposed to one prior trauma. The remaining participants had been exposed to between two to seven previous traumas (*n* = 158) and one participant had been exposed to ten previous traumas. The most common type of prior traumatic exposure was loss (*n* = 198) followed by being threatened with a weapon (*n* = 66), accidents (*n* = 69), witnessing someone being in danger or injured (*n* =45), fire (*n* =45), violence (*n* = 41), shock over traumatic exposure happening to someone close to you (*n* =25), disaster (*n* = 9), sexual abuse (*n* = 6), rape (*n* = 3), childhood neglect (*n* = 3), physical abuse (*n* = 2), and war (*n* = 2).

A total of 224 participants (60.4 %) did not report any exposure to major life events during the past year, whereas 138 participants (37.2 %) had experienced one or two, and the remaining nine participants (2.4 %) reported three to four life events. The most common types of life events were: death in the family (*n* = 55), job change (n = 39), illness in the family (n = 36), moving home (*n* = 22), burglary (*n* = 17), and divorce (*n* = 15). The majority of the participants had previously been exposed to bank robbery (*n* = 235, 63.3 %) as indicated by the relatively high percentage of exposure to threats with a weapon. Ninety-two participants had been exposed to one prior robbery (24.8 %), 49 participants had been exposed to two robberies (13.2 %), and the rest had been exposed to 3-10 previous robberies. Furthermore, one participant had experienced 14 robberies. Only 12 participants (3.2 %) feared that they were going to die during the robbery. Thus, fear of dying was excluded from further analyses.

Table 1. Sample characteristics of all predictor variables, and PTSD severity (N = 371)

Pre-trauma variables			
Age	*M* = 42.11	*SD* = 12.47	*R* = 20-65
Female	61.5 %	(*n* = 228)	
Prior robberies	*M* =1.63	*SD* = 2.01	*R* = 0-14
Prior traumatic exposure	*M* = 0.54	*SD* = 0.00	*R* = 0-4
Life events	*M* = 0.54	*SD* = 0.79	*R* = 0-4
ASI	*M* = 10.11	*SD* =8.75	*R* = 0-39
Neuroticisme	*M* = 12.16	*SD* =7.93	*R* = 0-41
Peri-trauma variables			
Directly exposed (yes)	80.1 %	(*n* = 297)	
Perceived life threat (yes)	14.0 %	(*n* = 52)	
Perceived helplessness (yes)	37.2 %	(*n* = 138)	
Intense horror (yes)	34.5 %	(*n* = 128)	
TIS	*M* = 7.42	*SD* = 6.86	*R* = 0-24
PRS	*M* = 9.78	*SD* = 6.00	*R* = 0-62
PDEQ	*M* = 17.12	*SD* = 7.26	*R* = 10-45
Post-trauma variables			
ASDS	*M* =34.84	*SD* =13.16	*R* = 9-86
PTCI NCS	*M* =1.53	*SD* = 0.63	*R* = 21-104
PTCI NCW	*M* =2.31	*SD* = 1.12	*R* = 7-43
PTCI SB	*M* = 1.86	*SD* = 1.00	*R* = 5-27
CSS POS	*M* = 29.55	*SD* = 4.76	*R* = 13-35
CSS NEG	*M* = 2.02	*SD* = 1.74	*R* = 1-7
CSS SATIS	*M* = 6.40	*SD* = 1.17	*R* = 1-7
CSQ RAT	*M* = 27.68	*SD* = 4.85	*R* = 15-42
CSQ EMO	*M* = 14.06	*SD* = 3.55	*R* = 10-30
CSQ AVO	*M* = 19.09	*SD* = 3.97	*R* = 10-32
Dependent variable			
HTQ	*M* = 25.12	*SD* = 7.94	*R* = 16-62

Note. PTSD (posttraumatic stress disorder), *M* (mean), *SD* (standard deviation), *R* (range), age (*n* = 370), ASI (Anxiety Sensitivity Index), TIS (Tonic Immobility Scale), PRS (Physical Reaction Scale), PDEQ (Peritraumatic Dissociation Scale), ASDS (Acute Stress Disorder Scale), PTCI (Posttraumatic Cognition Inventory), NCS (negative cognitions about self), NCW (negative cognitions about the world), SB (self-blame), CSS (Crisis Support Scale), POS (positive), NEG (negative), SATIS (satisfaction), CSQ (Coping Style Questionnaire), RAT (rational), EMO (emotional), AVO (avoidant), and HTQ (Harvard Trauma Questionnaire).

Predicting PTSD Severity

Table 2 shows the correlation matrix between the pre-trauma, peri-trauma, post-trauma predictor variables, and the HTQ total score. Of the pre-trauma variables female sex, anxiety sensitivity, and neuroticism were significantly associated with an increased HTQ total score (*rs* = .26-.45, *all ps* < .01). Age, prior bank robberies, prior traumatic exposures, and major life events were not significantly associated with the HTQ total score. All the peri-trauma variables with the exception of proximity to the robbery was significantly associated with an increased HTQ total score (*rs* = .28-.51, *all ps* < .01). The ASDS total score, avoidant coping, emotion-focused coping, negative social support, negative cognitions about self, negative cognitions about the world, and self-blame were all significantly associated with an increased score on the HTQ (*rs* = .22 -.64, *all ps* <.01). Furthermore, perceived social support was significantly associated with a decreased score on the HTQ ($r = -.14$, $p < .01$), whereas rational coping and social support satisfaction were not significantly associated with the HTQ total score.

Table 3 shows a hierarchical multiple regression analysis of the prediction of the HTQ total score from the pre-trauma, peri-trauma, and post-trauma predictor variables. The pre-trauma variables were entered into the model at the first step. The only significant factors were female sex, anxiety sensitivity and neuroticism. These factors accounted for approximately 28 % of the variance in the HTQ total score. At step two, the peritraumatic variables were entered. The model now accounted for 37 % of the variance explained in the HTQ total score with only prior bank robberies, anxiety sensitivity, neuroticism, and peritraumatic panic being significant factors. At the final step of the model, the posttraumatic variables were entered. The final model accounted for 51 % of the variance in the HTQ total score with only peritraumatic dissociation, ASDS total score, and negative cognitions about self remaining significant. The overall largest predictor of PTSD severity (i.e. the HTQ total score) was the ASDS total score ($\beta = .45$, $p < .001$). Thus, we investigated how much of the variance in the HTQ total score that could be explained alone by the ASDS total score. A linear regression analysis (not shown here) showed that the ASDS total score alone accounted for 41 % of the variance in the HTQ total score (adj. R^2=.41, $F(1,369) = 258.74$, $p< .001$). Thus, the inclusions of the 23 other predictor variables only improved the model by 10 %. Negative cognitions about self were the second largest predictor of PTSD severity ($\beta = .21$, $p < .001$).

Table 2. Correlations between pre-, peri-, post-trauma predictor variables and the HTQ total score (Pearson's r)

Measure	1	2	3	4	5	6	7	8	9	10	11	12	13	14	15	16	17	18	19	20	21	22	23	24
Age	-																							
Sex	-.09	-																						
Prior robberies	.36**	.02	-																					
Prior traumas	.08	-.03	.14**	-																				
Life events	-.17**	-.10	-.01	.16**	-																			
ASI	-.04	-.09	.00	-.04	.08	-																		
Neuro	-.17	-.29**	-.11*	.03	.13**	.34**	-																	
Proximity	-.04	-.12*	-.11*	.01	-.03	.08	.04	-																
Life threat	.13*	-.12*	.02	-.02	.03	.16**	.15**	.18**	-															
Helpless-ness	-.12*	-.19*	-.07	-.01	-.03	.27**	.23**	.09	.19**	-														
Horror	-.03	-.30**	-.09	-.08	.05	.28**	.26**	.12**	.25**	.44**	-													
TIS	-.09	-.23**	-.13	-.04	.03	.34**	.31**	.29**	.27**	.48**	.41**	-												
PRS	-.09	-.31**	.04	-.04	.12*	.48**	.37**	.18**	.37**	.48**	.52**	.52**	-											
PDEQ	-.09	-.26**	-.10	.08	.04	.37**	.32**	.18**	.27**	.43**	.37**	.49**	.64**	-										
ASDS	-.13*	-.30**	-.03	-.05	.15**	.48**	.36**	.09	.26**	.45**	.48**	.46**	.67**	.68**	-									
PTCI NCS	-.12*	-.22**	-.07	.05	.10	.42**	.58**	.11*	.27**	.29**	.26**	.37**	.49**	.44**	.58**	-								
PTCI NCW	-.06	-.17**	-.04	.01	.08	.38**	.42**	.01	.27**	.24**	.23**	.29**	.43**	.37**	.53**	.63**	-							
PTCI SB	-.07	.07	-.09	-.05	.08	.20**	.11*	.16**	.20**	-.07	.01	.01	.17**	.12*	.23**	.38**	.36**	-						
CSS POS	.10*	-.12*	.07	.01	-.02	-.12*	-.15**	-.03	.01	-.06	-.04	-.11*	-.08	-.13*	-.18**	-.26**	-.15**	-.10*	-					
CSS NEG	-.06	-.12	-.04	.00	.13*	.16**	.19**	.06	.01	.19**	.15**	.15**	.29**	.15**	.27**	.26**	.22**	.07	-.26**	-				
CSS SATIS	.07	.05	.05	-.05	-.02	-.11*	-.13*	.04	-.04	-.07	-.06	-.09	-.06	-.04	-.17**	-.16*	-.10	-.02	-.29**	-.23**	-			
CSQ RAT	.09	.07	.01	.00	-.07	.00	-.28**	.07	-.07	-.05	-.06	-.05	-.00	.04	.05	-.16**	-.02	.03	.07	-.00	.09	-		
CSQ EMO	-.12*	-.30**	-.07	.10	.12*	.39**	.71**	-.00	.09	.17**	.18**	.26**	.34**	.27**	.36**	.66**	.47**	.17**	-.14**	.22**	-.18**	-.24**	-	
CSQ AVO	.07	-.23**	.00	-.08	.02	.19**	.29**	.02	.17**	.17**	.20**	.21**	.28**	.31**	.33**	.33**	.30**	.09	.00	.15**	-.02	.16**	.31**	-
HTQ	-.07	-.26**	.03	.02	.07	.38**	.45**	.08	.28**	.36**	.36**	.38**	.51**	.41**	.64**	.61**	.50**	.22**	-.14**	.22**	-.10	-.07	.47**	.29**

Note. 1=men, 0= women, PTSD (posttraumatic stress disorder), *M* (mean), *SD* (standard deviation), *R* (range), age (n = 370), Neuro (neuroticism), proximity (proximity to the robbery), ASI (Anxiety Sensitivity Index), TIS (Tonic Immobility Scale), PRS (Physical Reaction Scale), PDEQ (Peritraumatic Dissociation Scale), ASDS (Acute Stress Disorder Scale), PTCI (Posttraumatic Cognition Inventory), NCS (negative cognitions about self), NCW (negative cognitions about the world), SB (self-blame), CSS (Crisis Support Scale), POS (positive), NEG (negative), SATIS (satisfaction), CSQ (Coping Style Questionnaire), RAT (rational), EMO (emotional), and AVO (avoidant).

* $p = \leq .05$, ** $p = \leq .01$.

Table 3. Hierarchical Multiple Regression Analysis Predicting the HTQ total score

	Step1 (β)	Step2 (β)	Step3 (β)
Pre-trauma variables			
Age	-.05	-.05	-.01
Gender	-.15**	-.06	-.02
Prior robberies	.08	.10*	.06
Prior traumatic exposure	.01	.02	.04
Life changes	-.02	-.02	-.06
ASI	.26***	.11*	.00
Neuroticism	.32***	.25***	.10
Peri-trauma variables			
Proximity to the robbery		-.02	.00
Perceived life threat		.09	.07
Perceived helplessness		-.07	.05
Perceived horror		.05	.02
TIS		.05	.03
PRS		.22***	.06
PDEQ		.05	-.15**
Post-trauma variables			
ASDS			.45***
PTCI NCS			.21**
PTCI NSW			.04
PTCI SB			.01
CSS POS			.04
CSS NEG			.01
CSS SATIS			.05
CSQ RAT			.00
CSQ EMO			.09
CSQ AVO			.01
adj. R^2	.28***	.37***	.51***
F(df)	$F(7,362) = 21.19$	$F^{change}(7,355) = 8.62$	$F^{change}(10,345) = 11.38$

Note. 1=men, 0= women, HTQ (Harvard Trauma Questionnaire), *M* (mean), *SD* (standard deviation), *R* (range), age (n = 370), ASI (Anxiety Sensitivity Index), TIS (Tonic Immobility Scale), PRS (Physical Reaction Scale), PDEQ (Peritraumatic Dissociation Scale), ASDS (Acute Stress Disorder Scale), PTCI (Posttraumatic Cognition Inventory), NCS (negative cognitions about self), NCW (negative cognitions about the world), SB (self-blame), CSS (Crisis Support Scale), POS (positive), NEG (negative), SATIS (satisfaction), CSQ (Coping Style Questionnaire), RAT (rational), EMO (emotional), and AVO (avoidant).

* $p \leq .05$, ** $p \leq .01$, *** $p \leq .001$.

Thus, we investigated how much of the variance in the HTQ total score could be explained alone by negative cognitions about self. A linear regression analysis (not shown here) showed that negative cognitions about self alone accounted for 37 % of the variance in the HTQ total score (adj. R^2=.37, F (1,369) = 216.64, $p< .001$). Thus, the inclusions of the 23 other predictor variables only improved the model by 14 %.

Discussion

The current study investigated the psychological impact of bank robbery on the employees by estimating the prevalence rates of both acute (i.e. ASD) and long-term post-traumatic stress symptoms (i.e. PTSD) in a cohort study of all Danish bank employees exposed to robbery during one year. At the same time, the current study also investigated several predictor variables grouped as pre-trauma, peri-trauma, and post-trauma predictors of PTSD severity. This knowledge can be used to facilitate preventive and early intervention which may hinder or minimize the development of PTSD and reduce the suffering and economic costs associated with PTSD.

Estimated ASD and PTSD Prevalence Rates

The prevalence rates of ASD (11.1 %) and PTSD (6.2 %) found in the current study are comparable to the prevalence rates found in our previous bank robbery study (ASD = 14.4 % and PTSD = 6.8 %; Hansen & Elklit, 2011; 2013). Thus, as expected, the rates are lower than those reported following other forms of robbery and nonsexual assault (ASD =17-25 % and PTSD = 20-30%; Brewin et al., 1999; Elklit, 1999; 2002; Elklit & Brink, 2004, Kleim et al., 2007). There are several possible explanations to why the reported prevalence rates of ASD and PTSD are lower following bank robbery than following other types of nonsexual assault. It has previously been argued that bank robberies may be less traumatizing than other forms of assault as violence during a bank robbery is often instrumental (i.e. directed toward achieving money) and not personal (i.e. directed toward the employees; Hansen & Elklit, 2013). However, the high subclinical prevalence rates of ASD and PTSD found in the current study indicate that it seems unlikely that exposure to bank robbery is simply less traumatizing than exposure to other forms of nonsexual assault. As argued by Hansen and Elklit (2013), bank employees are a special trauma population in relation to meeting the avoidance criteria. Bank employees have difficulty scoring high on avoidance as they have difficulty avoiding confrontations with the traumatic place, talking and thinking about the robbery as they usually return to work, talk to colleagues, participate in debriefing etc. This is also evident in the current study as both prevalence rates increase with the removal of the avoidance criterion (ASD without avoidance = 14.0 %, PTSD without avoidance= 18.1 %). Finally, the low and decreasing prevalence rates (11.1 % ASD to 6.2 % PTSD) could reflect normal remission or a positive effect of debriefing or social support.

Predictors of PTSD Severity

Pre-Trauma Predictor Variables

Contrary to our expectations, age, prior robberies, prior traumatic exposures, and exposure to major life events were not significantly associated with PTSD severity as measured by the HTQ total score. The results are somewhat in accordance with the Brewin et al. (2000) meta-analysis of risk factors for PTSD finding that the effects of age and prior traumatic exposure are small. Additionally, age, prior robberies, prior traumatic exposure, and exposure to major life events do not seem to have had a distal effect on PTSD severity either; as the bivariate correlations between these pre-trauma predictor variables and the other predictor variables were primarily non-significant or weak. It is possible that prior traumatic exposure and exposure to major life events may simply not be significant predictor of PTSD severity; as the most common forms of traumatic exposure and life events were robbery and loss of a close relative. As mentioned earlier, all types of traumatic exposure are not necessarily equal in their effect (Ozer et al., 2003). For instance exposure to sexual assault seems to be more traumatizing than experiencing loss or exposure to bank robbery (Frans, Rimmö, Åberg, & Frederikson, 2005). Thus, as we previously argued it appears that some bank robbery victims get used to being exposed to robberies and that prior robberies therefore may have a stress inoculation effect in these individuals (Hansen & Elklit, 2011). However, it is also possible that prior robbery exposure increases the risk of developing PTSD in some individuals, but not in others, and that these mixed results produce a non-significant relationships between prior exposure to bank robberies and PTSD severity.

As expected, female sex, anxiety sensitivity (i.e. negative cognitions about bodily sensations), and neuroticism were significantly associated with PTSD severity. Furthermore, all three factors were significant predictors of PTSD severity when controlled for the effect of the other pre-trauma predictor variables. However, the bivariate correlations were only weak to moderate and anxiety sensitivity and neuroticism did not significantly predict PTSD severity when controlled for the effect of post-trauma factors. Sex stopped being a significant predictor already when controlled for the effect of the peri-trauma factors. The relationship between sex and PTSD is likely to be much more complex than what the current study suggests. According to Christiansen and Elklit (2012) there are sex differences in several risk factors associated with the development of PTSD. Thus, sex may have an indirect influence on the development of PTSD, which is mediated by other factors risk factors. Both anxiety sensitivity and neuroticism were significantly and moderately to strongly associated with negative cognitions about self (rs = .42-.58, $ps < .01$) and the ASDS total score (rs = .36-48, $ps < .01$), both significant predictors of PTSD severity at the final step of the regression analysis. This suggests that anxiety sensitivity and neuroticism did not contribute anything unique when these two posttraumatic variables were introduced. Thus, negative cognitions about bodily sensations seem less important than negative cognitions about self in the genesis of PTSD. This is in accordance with research showing that negative cognitions about self seem to have a more pronounced role in the development of PTSD than other forms of negative cognitions (Karl et al., 2009; O'Donnell, Elliott, Wolfgang, & Creamer, 2007). Furthermore, it is possible that neuroticism was only a significant predictor of PTSD in the first place due to the cognitive element in neuroticism or that the effect of neuroticisme on PTSD is mediated by negative cognitions about self.

Peri-Trauma Predictor Variables

As expected perceived life threat, perceived helplessness, intense horror, tonic immobility, peritraumatic panic, and peritraumatic dissociation were all positively associated with PTSD severity. The associations were mainly moderate (*all rs* =.28-.51), and only peritraumatic dissociation was a significant predictor of PTSD severity when controlled for the effect of the other predictor variables. However, it is not possible to assess whether or not peritraumatic dissociation is truly a significant predictor of PTSD severity in the current study. The bivariate correlation between peritraumatic dissociation and PTSD severity indicates that peritraumatic dissociation is a predictor of PTSD, which is in accordance with previous research (Breh & Seidler, 2007; Ozer et al., 2003). However, in the complete model, PDEQ served as a negative suppressor variable (Conger, 1974) for another variable, as indicated by further analyses (not shown here) most probably ASD. That is, it removed variance of the variable ASD irrelevant for the prediction of PTSD severity. By this statistical suppression effect, the regression weight becomes opposite in sigh to what is to be expected. As shown in table 2, the bivariate correlation of PDEQ and PTSD was .41. Thus, it is not possible to be sure whether or not peritraumatic dissociation is a significant predictor of PTSD in the current study due to the suppressor effect. However, the relationship between peritraumatic dissociation and PTSD seems rather complex and dependent of ASD in the current study as shown in the van der Velden and Wittmann (2008) review.

In accordance with previous bank robbery studies and studies of other forms of traumatic exposure (Hansen & Elklit, 2013; Kleim et al., 1999; Miller-Burke et al., 1999; Ozer et al., 2003); perceived life threat, perceived helplessness and intense horror were bivariately significantly associated with increased PTSD severity. However, none of these three variables predicted PTSD severity when controlling for the effect of the other predictor variables. Perceived life threat, perceived helplessness, and intense horror correlated significantly with ASD severity and negative cognitions about self, both of which were significant predictors of PTSD severity when all other factors were controlled for. Thus, perceived life threat, perceived helplessness, and horror could have had an indirect effect on PTSD through either or both of these two other factors. However, it is more likely that perceived life threat is simply not a particularly strong risk factor of PTSD severity in the current study as indicated by the weak association between perceived life threat and PTSD severity (r =.28) and the fact that only 14 % of the participants (n= 52) reported perceived life threat. The results regarding perceived helplessness and horror are in accordance with our previous bank robbery study (Hansen & Elklit, 2013), the Friedman et al. (2010) review of the diagnostic criteria for the DSM-IV-R PTSD diagnosis and the newly introduced PTSD diagnosis in the DSM-5 (APA, 2013). We previously found that perceived helplessness was moderately associated with PTSD severity but not a significant predictor of PTSD severity when the effect of other predictor variables was controlled for (Hansen & Elklit, 2013). Furthermore, according to Friedman et al.'s (2010) review, the A2 criterion has limited capacity to predict PTSD and does not improve diagnostic accuracy, as research shows that some people develop PTSD without an initial A2 response. Thus, the A2 criterion has been removed from the PTSD diagnostic criteria in the DSM-5 (APA, 2013). The A2 criterion also seems to have limited utility in the prediction of PTSD severity following bank robbery.

In accordance with previous research, tonic immobility and peritraumatic panic were significantly and moderately associated with PTSD severity (Bryant et al., 2011; Bryant & Panasetis, 2001; Heidt et al., 2005; Kunst et al., 2011; Marx et al., 2008; Nixon & Bryant, 2003; Rocha-Rego et al., 2009). However, tonic immobility and peritraumatic panic were not significant predictors of PTSD severity, when the effect of the other variables was controlled for. Tonic immobility and peritraumatic panic were moderately and significantly associated with increased ASD severity and negative cognitions about self. Thus, tonic immobility and peritraumatic panic might still have had an indirect effect on the development of PTSD severity through either or both of these factors. The indirect effect of peritraumatic panic is especially likely to have occurred through ASD severity, as the association between peritraumatic panic and ASD severity was particularly strong. The association between peritraumatic panic and peritraumatic dissociation was also strong and peritraumatic dissociation has been found to mediate the effect of panic on PTSD (Bryant et al., 2011). Although, it is not possible to assess whether or not peritraumatic dissociation is a significant predictor of PTSD severity in the current study, the effect of peritraumatic panic does not seem to be affected by peritraumatic dissociation in the current study, as peritraumatic panic was a significant predictor of PTSD severity when controlled for peritraumatic dissociation. In relation to tonic immobility the Kunst et al. (2011) study also found that tonic immobility was not a significant risk factor for PTSD following assault when controlled for other peritraumatic factors. At the same time, it has been argued that the impact of tonic immobility on PTSD could depend on a third factor for instance self-blame (cf. Heidt et al., 2005; Kunst et al., 2007). Thus, tonic immobility may only be strongly associated with PTSD if it leads to self-blame. This is not the case in the current study as tonic immobility did not correlate significantly with self-blame. Additionally, it is important to mention that tonic immobility following bank robbery may not be of the same clinical relevance as it seems to be following other forms of assault (cf. Heidt et al., 2005; Kunst et al., 2011). During a bank robbery the employees are often threatened to stand still, however, this does not equal not being able to move.

Post-Trauma Predictor Variables

As expected, ASD severity, negative cognitions about self, negative cognitions about the world, and self-blame were significantly associated with increased PTSD severity. Furthermore, ASD severity was the largest predictor of PTSD severity when the other predictor variables were controlled for. An additional regression analysis showed that ASD severity alone explained 41 % of the variance of PTSD severity and adding the other predictor variables to the model only increased the amount of PTSD severity accounted for by 10 %. In other words, ASD severity plays a pivotal role in the prediction of PTSD following bank robbery. This is in contrast to previous studies showing that theoretically derived cognitive factors seem to be better predictors of PTSD than ASD (Kleim et al., 2007). Although, the ASD diagnosis may have limited utility in the prediction of PTSD diagnosis (Bryant, 2011), the results indicate that ASD severity seem to a strong predictor of PTSD severity following bank robbery. This may indicate that although there are different diagnostic criteria for ASD and PTSD, it is possible that ASD severity is an early indication of PTSD severity. This is in accordance with studies showing that early posttraumatic symptoms are a strong predictor of later posttraumatic symptoms (cf. Brewin et al., 1999; Hansen & Elklit, 2013; O'Donnell et al., 2007). A special focus on the employees' ASD severity levels in the acute phrase of the robbery may help to identify many victims at risk of developing PTSD. We therefore recommend the use of the ASDS to screen for PTSD in the acute phrase following bank robbery exposure. Cognitive factors and in particular negative cognitions about self also seem to play a pronounced role in the development of PTSD severity following bank robbery. Thus, the results of our study are in accordance with the O'Donnell et al. (2007) study, which also found that posttraumatic cognitions make a unique contribution to explaining the development of posttraumatic symptoms over and above that of early posttraumatic symptoms. However, this was only in relation to negative cognitions about self in the current study.

Negative cognitions about self were the second largest predictor of PTSD severity in the multiple regression analysis. Furthermore, an additional regression analysis showed that negative cognitions about self alone explained 37 % of the variance of PTSD severity, thus adding the other predictor variables to the model only increased the amount of PTSD severity accounted for by 14 %. However, negative cognitions about the world and self-blame were not significant predictor of PTSD severity when the effect of the predictor variables was controlled for. The association between negative cognitions about the world and PTSD severity was strong, but the association between PTSD severity and self-blame was only weak. Negative cognitions about the world were strongly associated with both negative cognitions about self and ASD severity, which may explain why negative cognitions about the world was not a predictor of PTSD severity despite the strong association. In other words, negative cognitions about the world do not seem to add anything unique on its own when controlled for negative cognitions about self and ASD severity. Furthermore, self-blame has been found to be both positively, negatively and not associated with PTSD severity following different types of traumatic exposure (Blain, Galovski, Elwood, & Meriac, 2012; Karl et al., 2009; O'Donnell et al., 2007). The discrepancies in the results may be due to self-blame becoming protective when it is linked to increased sense of control (i.e. behavioral self-blame) rather than to self-esteem (i.e. characterlogical self-blame; Blain et al., 2012; Karl et al., 2009). Thus, in the current study it is possible that self-blame is only weakly associated with PTSD severity as it is may be a protective factor for some of the employees, a risk factor for other employees or/and not significantly associated with PTSD severity in others. Instead, the employees' negative cognitions about self seem particularly important to the development of PTSD. Negative cognitions about self are also found to have a more pronounced role than negative cognitions about self and self-blame in the development of PTSD following other forms of traumatic exposure (Blain et al., 2012; Karl et al. 2009; O'Donnell et al., 2007). Furthermore, we only investigated acute posttraumatic cognitions. However, research shows that negative cognitions about self become even more influential than negative cognitions about the world in the development of PTSD over time (O'Donnell et al., 2007). Thus, an internally driven sense of threat (i.e. negative cognitions about self) seems to be a more power full mechanism in the development of PTSD than an external sense of threat (i.e. negative cognitions about the world) following bank robbery as found in the O'Donnell et al. (2007) study of posttraumatic appraisals in the development of posttraumatic stress symptoms in injury survivors. Even though a bank robbery is caused by human desperation, an external sense of threat seems less important than the internal. This suggests that it may be possible to take preventive actions against the development of PTSD following bank robbery by focusing on changing the way that employees think about themselves and their capacity to response to bank robbery. The importance of cognitions in the development of PTSD following bank robbery is not only in accordance with the Ehlers and Clark (2000) cognitive model of trauma response, but also the recently introduced PTSD diagnosis in the DSM-5; where negative cognitions constitute a forth symptom cluster (APA, 2013).

Contrary to our expectations, not all of the investigated measures of social support and coping were significantly associated with PTSD severity. Positive social support was significantly associated with decreased PTSD severity and negative social support, avoidant coping, and emotion-focused coping were significantly associated with increased PTSD severity. However, none of these post-trauma variables could predict PTSD severity when the other predictor variables were controlled for. Furthermore, rational coping and social support satisfaction were not significantly associated with PTSD severity. Rational coping has also been found unassociated with PTSD following other forms of traumatic exposure (Bödvarsdóttir & Elklit, 2004), which may suggest that it is not so much the absence of rational problem-focused coping that is associated with PTSD but rather the presence of more maladaptive coping styles. The different measures of social support may not be so strongly associated with PTSD severity in the current study because social support was measured one week after the robbery. The association between social support and posttraumatic symptoms has been shown to increase over time (Richards, 2000; Ozer et al., 2003). Thus, it is possible that social support will become a stronger predictor of PTSD severity following bank robbery over time. Furthermore, the lack of a stronger association between the different types of social support and PTSD severity may be due to the use of the CSS to measure social support. The CSS and especially the use of single items may not be sensitive enough to capture the range of social support that the participants encountered. Although, emotion-focused coping was moderately associated with PTSD severity (r = .47), emotion-focused coping was not a significant predictor of PTSD severity when controlled for the other predictor variables. It has been argued that many coping measures, including the CSQ, confound emotion-focused coping with an emotional outcome, specifically distress (cf. Christiansen & Elklit, 2011; Stanton, Danoff-Burg, Cameron, & Andrews, 1994). Thus, emotion-focused coping was likely not to be a significant predictor in the current study as we controlled for the effect of neuroticism. This assumption is supported by the strong association between emotion-focused coping and neuroticism (r =.71). The use of another coping scale may have produced a different result. However, many coping scales exhibit this confounding problem (Stanton et al., 1994).

LIMITATIONS

The current study has several limitations. Firstly, both the ASD and the PTSD prevalence rates were based on self-report measures rather than clinical interviews. Although both ASD and PTSD were assessed with standardized scales and according to the DSM-IV-TR, the estimated ASD and PTSD prevalence rates may still be imprecise. Secondly, although the translation and back-translation of several of the scales into Danish were done independently, the psychometric properties of the original English scales cannot necessarily be transferred to the Danish versions. Thirdly, although there were no significant differences in ASD severity and PTSD severity between the debriefed and non-debriefed participants, debriefing may still have affected the results. It is possible that debriefing may have affected some of the participants' recovery process negatively and maybe other participants' recovery process positively. Finally, it is possible that the single items used to measure negative social support and social support satisfaction are not sensitive enough to capture the complexity of these aspects of social support that the participants experienced after the robbery.

CONCLUSION

Despite these limitations the current study is the only national cohort bank robbery study investigating the psychological impact and predictors of PTSD following bank robbery ever conducted. Among the strengths of the current study is a large and highly representative sample of bank robbery victims including a high initial response rate and a low dropout rate. The results show that exposure to bank robbery is a potentially traumatic event. The ASD prevalence rate was 11.1 % and subclinical ASD rates ranged from 11.1 % to 26.7 %. The PTSD prevalence rate was 6.2 % and subclinical PTSD prevalence rates ranged from 6.2 % to 18.1 %. The high subclinical prevalence rates indicate that many participants, who do not meet the diagnostic criteria for a full diagnosis, still suffer from severe symptoms of both ASD and PTSD.

The current study investigated a unique combination of predictors of PTSD severity in the attempt to identify and understand the psychological factors involved in the complex relationship between exposure to bank robbery and the development of PTSD. Results showed that 51 % of the variance in PTSD severity could be explained with only peritraumatic dissociation, ASD severity, and negative cognitions about self being significant predictors. However, due to a statistical artifact it was not possible to assess how peritraumatic dissociation is related to PTSD severity in the multivariate analysis. ASD severity and negative cognitions about self are of particular importance in the development of PTSD following bank robbery as they alone explained 41 % and 37 % of the variance in PTSD severity, respectively. Thus, victims with high levels of ASD severity are at increased risk of developing PTSD and screening for PTSD following bank robbery should focus on ASD severity. At the same time, in accordance with a cognitive model of PTSD and the DSM-5, posttraumatic negative cognitions about self also play a pivotal role in the development of PTSD symptoms following bank robbery. Thus, future research should focus on the possibilities of developing preventive actions against the development of PTSD by focusing on negative cognitions.

REFERENCES

Agaibi, C. E., & Wilson, J. P. (2005). Trauma, PTSD and Resilience – a Review of the Literature. *Trauma, Violence & Abuse*, *6*, 195-216. doi: 10.1177/1524838005277438.

Alexander, D.A., & Wells, A. (1991). Reactions of police officers to body-handling after a major disaster. A before-and-after comparison. *British Journal of Psychiatry*, *159*, 547–555.

Andrews, B., Brewin, C. R., & Rose, S. (2003). Gender, social support and PTSD in victims of violent crime. *Journal of traumatic stress*, *16*, 421-427. doi: 10.1023/A:1024478305142.

American Psychiatric Association (1994). *Diagnostic and Statistical Manual of Mental Disorders, 4th ed.* Washington DC: American Psychiatric Association.

American Psychiatric Association (2000). *Diagnostic and statistical manual of mental disorders: DSM-IV-TR*, *4th ed*, text revision. Washington, DC: Author.

American Psychiatric Association (2013). *Diagnostic and Statistical Manual of Mental Disorders, 5th ed.* Washington DC: American Psychiatric Association.

Armour, C., Elklit, A., & Shevlin, M. (2011). The latent structure of acute stress disorder: a posttraumatic stress disorder approach. *Psychological Trauma: Theory, Research, Practice, and Policy, 5,*18-25. doi: 10.1037/a0024848.

Bach, M. (2003). En empirisk belysning og analyse af "Emotionel Numbing" som eventuel selvstændig faktor i PTSD. [An empirical illumination and analysis of emotional numbing as an independent factor in PTSD]. *Psykologisk studieskriftserie, 6*, 1-199.

Beck, G. J., Coffey, S. F., Palyo, S. A., Gudmundsdottir, B., Miller, L. M., & Colder, G. R. (2004). Psychometric properties of the posttraumatic cognitions inventory (PTCI): a replication with motor vehicle accident survivors. *Psychological Assessment 16*, 289-298. doi:10.1037/1040-3590.16.3.289

Birmes, P., Brunet, A., Benoit, M., Defer, S., Hatton, L., Sztulman, H., et al. (2005). *Validation of the Peritraumatic Dissociative Experiences Questionnaire self report.*

version in two samples of French-speaking individuals exposed to trauma. *European Psychiatry, 20*, 145-151. doi : 10.1016/j.eurpsy.2004.06.033.

Blain, L. M., Galovski, T., Elwood, L. S., & Meriac, J. P. (2012). How does the posttraumatic cognitions inventory fit in a four-factor posttraumatic stress disorder world? An initial analysis. *Psychological Trauma: Theory, Research, Practice, and Policy*. doi:10.1037/a0030251.

Breh, D. C., & Seidler, G. H. (2007). Is peritraumatic dissociation a risk factor of PTSD? *Journal of Trauma and Dissociation, 8,* 53-69. doi: 10.1300/J229v08n01_04.

Breslau, N., & Schultz, L. (2013). Neuroticism and post-traumatic stress disorder: a prospective investigation. *Psychological Medicine, 43*, 1697-1702. doi:10.1017/S0033291712002632.

Brewin, C., Andrews, B., & Rose, S. (2000). Fear, helplessness, and horror in posttraumatic stress disorder: Investigating *DSM-IV* Criterion A2 in victims of violent crime. *Journal of Traumatic Stress, 13,* 499-509. doi: 10.1023/A:1007741526169.

Brewin, C. R., Andrews, B., Rose, S., & Kirk, M. (1999). Acute Stress Disorder and Posttraumatic Stress Disorder in Victims of Violent Crime. *American Journal of Psychiatry, 156*, 360-366. Retrieved from http://journals.psychiatryonline.org/data/Journals/AJP/3698/360.pdf.

Brewin, C. R., Andrews, B., & Valentine, J. D. (2000). Meta-analysis of risk factors for posttraumatic stress disorder in trauma-exposed adults. *Journal of Consulting and Clinical Psychology*, *68*, 748-766. doi: 10.1037//0022-006x.68.5.748.

Briggs, S. R., & Cheek, J. M. (1986). The role of factor analysis in the development and evaluation of personality scales. *Journal of Personality, 54,* 106-148. doi: 10.1111/j.1467-494.1986.tb00391.x.

Bryant, R. A. (2003). Early Predictors of Posttraumatic Stress Disorder. *Biological Psychiatry*, *53*, 789-795. doi:10.1016/S0006-3223(02)01895-4.

Bryant, R. A. (2009). Is peritraumatic dissociation always pathological? In Dell, P. F., & O'Neil, J. A. *Dissociation and the dissociative disorders – DSM-V and beyond.* New York and London: Routledge Taylor and Francis Group.

Bryant, R. A. (2011). Acute stress disorder as a predictor of posttraumatic stress disorder. *Journal of Clinical Psychiatry, 72*, 233-239. doi: 10.4088/JCP.09r05072blu.

Bryant, R. A., Brooks, R., Silove, D., Creamer, M., O'Donnell, M., & McFarlane, A. C. (2011). Peritraumatic dissociation mediates the relationship between acute panic and chronic posttraumatic stress disorder. *Behaviour Research and Therapy, 49,* 346-351.

Bryant, R., & Harvey, A. (1997). Acute stress disorder: a critical review of diagnostic issues. *Clinical Psychology Review, 17*, 757-773. doi: 10.1016/S0272-7358(97)00052-4.

Bryant, R. A., Moulds, M. L., & Guthrie, R. M. (2000). Acute stress disorder scale. *Psychological Assessment, 12,* 61-68. doi: 10.1037//I040-3590.12.1.61.

Bryant, R. A., & Panasetis, P. (2001). Panic symptoms during trauma and acute stress disorder. *Behaviour Research and Therapy, 39,* 961-966. Retrieved from http://www.sciencedirect.com/science/article/pii/S000579 6700000863.

Bunting, B. P., Adamson, G., & Mulhall, P. (2002). A Monte Carlo examination of MTMM model with planned incomplete data structures. *Structural Equation Modeling*, *9*, 369–389. doi: 10.1207/S15328007SEM 0903_4.

Buunk, B. P., & Hoorens, V. (1992). Social support and stress. The role of social comparison and social exchange processes. *British Journal of Clinical Psychology*, *31*, 445-457. doi:10.1111/j.2044-8260.1992.tb 01018.x.

Bödvarsdóttir, I., & Elklit, A. (2004). Psychological reactions in Icelandic earthquake survivors. *Scandinavian Journal of Psychology, 45,* 3-13. doi: 10.1111/j.1467-9450.2004.00373.x.

Christiansen, D. & Elklit, A. (2008). Risk Factors predict post-traumatic stress disorder differently in men and women. *Annals of General Psychiatry, 7(24*). doi:10.1186/1744-859X-7-24.

Christiansen, D., & Elklit, A. (2012). Sex differences in PTSD. In E. Ovuga (Ed.). *Posttraumatic stress disorder in a global context* (pp.113-142). In Tech: open access book.

Conger, A. J. (1974). A revised definition for suppressor variables: a guide to their identification and interpretation. *Educational and Psychological Measurement, 34*, 35-46. doi: 10.1177/001316447403400105.

Costa, P. T., & McCrae, R. R. (1987). Neuroticism, somatic complaints and disease: is the bark worse than the bite? *Journal of Personality, 55,* 299-316.

Costa, P. T., & McCrae, R. R. (2004). *NEO PI-R – clinical manual.* Danish translation, Virum: Hogrefe Psykologisk Forlag.

The Danish Bankers Association (2013). *Robbery statistics.* Retrieved September 24, 2013 from http://www.finansraadet.dk/tal--fakta/statistik-og-tal/roeveristatistik.aspx.

Ehlers, A., & Clark, D. M. (2000). A cognitive model of posttraumatic stress disorder. *Behaviour Research and Therapy, 38*, 319-345. doi: 10.1016/S0005-7967(99)00123-0.

Elklit, A. (1996). Coping styles questionnaire: a contribution to the validation of a scale for measurering coping strategies. *Personality and Individual differences, 21*, 809-812. Doi: 10.1016/0191-8869(96)00123-7.

Elklit, A. (1999). Røveriofre: De psykologiske konsekvenser af væbnet røveri. [Psychological consequences of armed robbery]. *Psykologisk Skriftserie, 24*(5), 1-75.

Elklit, A. (2002). Acute stress disorder in victims of robbery and victims of assault. *Journal of Interpersonal Violence, 17,* 872-887. doi:10.1177/0886260502017008005.

Elklit, A., & Brink, O. (2004). Acute stress disorder as a predictor of posttraumatic stress disorder in physical assault victims. *Journal of Interpersonal Violence, 19*, 709-726. doi:10.1177/0886260504263872.

Elklit, A., & Christiansen, D. (2010). ASD and PTSD in rape victims. *Journal of Interpersonal Violence, 25*, 1470-1488. doi: 10.1177/08862605093 54587.

Elklit, A., Pedersen, S. S., & Jind, L. (2001). The crisis support scale: Psychometric qualities and further validation. *Personality and Individual Differences, 31*, 1291-1302. doi:10.1016/S0191-8869(00)00220-8.

Falsetti, S.A., & Resnick, H.S. (1992). *The physical reaction scale.* Charleston, SC: Crime Victims Research and Treatment center, Medical University of South Carolina.

Falsetti, S. A., Resnick, H. S., Dansky, B. S., Lydiard, R. B., & Kilpatrick, D. G. (1995). The relationship of stress to panic disorder: cause and effect. In Mazure, C. M. *Does stress cause psychiatric illness?* Washington, DC: American Psychiatric Press, 111-147.

Frans, Ö., Rimmö, P. A., Åberg, L., & Frederikson M. (2005). Trauma exposure and post-traumatic stress disorder in the general population. *Acta Psychiatrica Scandinavica, 111*, 291-299. doi: 10.1177/104365960629 1549.

Friedman, M. J., Resick, P.A., Bryant, R. A., & Brewin, C. (2010). Considering PTSD for DSM-5. *Depression & Anxiety,* 2011; 28: 750-769. doi: 10.1002/da.20767.

Foa, E. B., Ehlers, A., Clark, D. M., Tolin, D. F., & Orsillo, S. M. (1999). The Posttraumatic Cognitions Inventory (PTCI): Development and validation. *Psychological Assessment, 11*, 303-314. doi: 10.1037/1040-3590.11.3.303.

Gallup, G. G. (1977). Tonic immobility: the role of fear and predation. *Psychological Record, 27,* 47-61.

Gilioli, R., Campanini, P., Fichera, G. P., Punzi, S., & Gassitto, M. G. (2006). Emerging aspects of psychosocial risks: violence and harassment at work. *La Medicina del Lavoro, 97,* 160-164.

Hansen, M., & Elklit, A. (2011). Predictors of acute stress disorder in response to bank robbery. *European Journal of Psychotraumatology, 2.* doi: 10.3402) ejpt.v2i0.5864.

Hansen, M., & Elklit, A. (2013). Does acute stress disorder predict posttraumatic stress disorder following bank robbery? *Journal of Interpersonal Violence, 28*, 25-44. doi: 10.1177/0886260512448848.

Heidt, J., Marx, B. P., & Forsyth, J. P. (2005). Tonic immobility and childhood sexual abuse: a preliminary report evaluating the sequelae of rape-induced paralysis. *Behavior Research and Therapy, 43,* 1157-1171. Retrieved from http://www.sciencedirect.com/science/article/pii/S0005796704002220.

Holeva, V., Tarrier, N., & Wells, A. (2001). Prevalence and Predictors of Acute Stress Disorder and PTSD Following Road Traffic Accidents: Thought Control Strategies and Social Support. *Behaviour Therapy, 32*, 65-83. Retrieved from http://www.sciencedirect.com/science/article/pii/S0005789401800447.

Joseph, S., Andrews, B., Williams, R., & Yule, W. (1992) Crisis support and psychiatric symptomatology in adult survivors of the Jupiter cruise ship disaster. *British Journal of Clinical Psychology, 31*, 63-73. doi:10.1111/j.2044-8260.1992.tb00968.x.

Kamphuis, J. H., & Emmelkamp, P. M. G. (1998). Crime-Related Trauma: Psychological Distress in Victims of Bank robbery. *Journal of Anxiety Disorders,12*, 199-208. doi:10.1016/SO887-6185.

Karl, A., Rabe, S. Zöllner, T., Maercker, A., & Stopa, L. (2009). Negative self-appraisals in treatment-seeking survivors of motor vehicle accidents. *Journal of Anxiety Disorders,23*, 775-781. doi:10.1016/j.janxdis.2009.03.001.

Kessler, R.C., Sonnega, A., Bromet, E., Hughes, M. & Nelson, C. B. (1995). Posttraumatic Stress Disorder in the National Comorbidity Survey. *Archives of General Psychiatry*, *52*, 1048-1060. doi: 10.1001/archpsy.1995.03950240066012.

Kleim, B., Ehlers, A., & Glucksman, E. (2007). Early predictors of chronic post-traumatic stress disorder in assault survivors. *Psychological Medicine, 37*, 1457-1467. doi:10.1017/S0033291707001006.

Kunst, M., Winkel, F. W., & Bogaerts, S. (2011). Recalled Peritraumatic Reactions, Self-Reported PTSD, and the Impact of Malingering and Fantasy Proneness in Victims of Interpersonal Violence Who Have Applied for State Compensation. *Journal of Interpersonal Violence, 26,* 2186-2210. doi: 10.1177/0886260510383032.

Ladwig, K.H., Marten-Mittag, B., Dammann, G., Erazo, N., Reicherzer, M. Baedeker, U., & Stolzewski, B. (2002). Prädiktoren persistierender psychischer morbidität bei bankmitarbeitern nach räuberischen überfällen auf Bank-Institute. [Predictors of persistent affective disability in bank employees exposed to violent bank robberies]. *Zeitschrift für Medizinische Psychologie, 3*, 105-111. Retrieved from "http://iospress.metapress.com/content/110621".

Lazarus, R., & Folkman, S. (1984). *Stress, appraisal and coping*. New York: Springer.

Litz, B. T., & Maguen, S. (2007). Early Intervention for Trauma. 540-561. I Friedman, M. J., Keane, T. M., & Resick, P. (2007) (eds.). *Handbook of PTSD – Science and Practice*. New York & London: Guilford Press.

Marmar, C.R., Weiss, D. S., & Metzler, T. J. (2004). The peritraumatic dissociative experiences questionnaire. I Wilson, J. P., & Keane, T. M. (2004) (eds.). *Assessing psychological trauma and PTSD.* New York & London: Guilford Press.

Marx, B.P., Forsyth, J.P., Gallup, G.G., Fusé, T., & Lexington, J.M. (2008).Tonic Immobility as an Evolved Predator Defense: Implications for Sexual Assault Survivors. *Clinical Psychology: Science and Practice, 1,* 74-90. doi: 10.1111/j.1468-2850.2008.00112.x.

Miller-Burke, J., Attridge, M., & Fass, P. M. (1999). Impact of traumatic events and organizational response. *Journal of Occupational and Environmental Medicine,41*, 73-83. doi:10.1097/00043764-199902000-00001.

Mollica, R. F., Caspi-Yavin, Y., Bollini, P., Truong, T., Tor, S., & Lavelle, J. (1992). Harvard Trauma Questionnaire. Validating a cross-cultural instrument for measuring torture, trauma, and posttraumatic stress disorder in Indochinese refugees. *Journal of Nervous and Mental Disease, 180,* 111-116.

Monat, A., & Lazarus, R. S. (1991). Stress and coping: An anthology. New York: Columbia University Press.

Nixon, R. D., & Bryant, R. A. (2003). Peritraumatic and persistent panic attacks in acute stress disorder. *Behavior Research and Therapy, 41*, 1237-1242. Retrieved from http://www.sciencedirect.com/science/article/pii/S0005796703001505.

Nixon, R. D., & Bryant, R. A. (2005). Are negative cognitions associated with severe acute trauma responses? *Behavior Change, 22,* 22-28. doi: http://dx.doi.org/10.1375/bech.22.1.22.66783.

O'Donnell, M. L., Elliott, P., Wolfgang, B. J., & Creamer, M. (2007). Posttraumatic appraisals in the development and persistence of posttraumatic stress symptoms. *Journal of Traumatic Stress, 20*, 173-182. doi:10.1002/jts.20198.

Ozer, E. J., Best, S. R., Lipsey, T. L., & Weiss, D. S. (2003). Predictors of posttraumatic stress disorder and symptoms in Adults: A meta-Analysis. *Psychological Bulletin, 129*, 52-73. doi: 10.1037/0033-2909.129.1.52.

Parslow R. A., Jorm, A. F., & Christensen, H. (2006). Association of pre-trauma attributes and trauma exposure with screening positive for PTSD: analysis of a community-based study of 2,085 young adults. *Psychological Medicine, 36*, 387–395. doi:http://dx.doi.org/10.1017/S0033291705006306.

Reiss, S., Peterson, R. A., Gursky, D. M., & McNally, R. J. (1986). Anxiety sensitivity. *Behaviour, Research and Therapy, 24,.1-8.* Retrieved from http://www.sciencedirect. com/science/article/pii/0005796786901439.

Richards, D. (2000). Symptom severity, personal and social variables after armed robbery. *British Journal of Clinical Psychology, 39*, 415-419. doi: 10.1348/014466500163419.

Rocha-Rego, V., Fiszman, A., Portugal, L. C., Pereia, M. G., de Oliveira, L., Mendlowicz, M. et al. (2009). Is tonic immobility the core sign among conventional peritraumatic signs and symptoms listed for PTSD? *Journal of Affective Disorders, 115,* 269-273. Retrieved from http://www.jad-journal.com/article/S0165-0327(08)00362-5/abstract.

Roger, D., Jarvis, G., & Narajin, B. (1993). Detachment and coping: the construction of a new scale for measuring coping strategies. *Personality and Individual Differences, 15*, 619-626. doi:10.1016/0191-8869(93)90003-L.

Rossier, J., Meyer de Stadelhofen, F., & Berthoud, S. (2004). The Hierarchical Structures of the NEO PI-R and the 16 PF 5*. *European Journal of Psychological Assessment, 20*, 27–38. doi:10.1027//1015-5759.20.1.27.

Schnider, K. R., Elhai, J. D., & Gray, M. J. (2007).Coping style use predicts posttraumatic stress and complicated grief symptom severity among college students reporting a traumatic loss. *Journal of Counseling Psychology, 54*, 344–350. doi: 10.1037/0022-0167.54.3.344.

Stanton, A. L., Kirk, S. B., Cameron, C. L., & Danoff-Burg, S. (2000). Coping through emotional approach: Scale construction and validation. *Journal of Personality and Social Psychology, 78,* 1150-1169. doi: 10.1037/0022-3514.78.6.1150.

Taylor, S., Koch, W. J., & McNally, R. J. (1992). How does anxiety sensitivity vary across the anxiety disorders? *Journal of Anxiety Disorders, 6*, 249-259.

van der Velden, P. G., & Wittmann, L. (2008). The independent predictive value of peritraumatic dissociation for PTSD symptomatology after type I trauma: a systematic review of prospective studies. *Clinical Psychology Review, 28,* 1009-1020. doi: 10.1016/j.cpr.2008.02.006.

Volchan, E., Souza, G.G., Franklin, C. M., Norte, C.E., Rocha-Rego, V., Oliverira, J. M. et al. (2011). Is there tonic immobility in humans? Biological evidence from victims of traumatic stress. *Biological Psychology, 88,* 13-19. Retrieved from http://www.sciencedirect.com/science/article/pii/S0301051111001463.

WHO (World Health Organization) (1992). The ICD-10 classification. Clinical descriptions and diagnostic guidelines. WHO: Genève.

Yap, M. B. H., & Devilly, G. J. (2004). The role of perceived social support in crime victimization. *Clinical Psychology Review, 24*, 1-14. doi: 10.1016/j.cpr.2003.09.007.

Reviewed by: Prof. Lutz Wittmann, PhD, International Psychoanalytic University Berlin, Germany & Dorte M. Christiansen, MSc, Institute for Psychology, University of Aarhus, Denmark.

INDEX

N

O

P

Q

R

S

T